"A compelling chronicle of the many, many (many) man-caused haz-ards that have threatened the largest source of accessible freshwater in the world."　　　　　—Susan Glaser, *Cleveland Plain Dealer*

"A marvelous work of nonfiction, which tells the story of humanity's interference with the natural workings of the world's largest unfrozen freshwater system."　　　　—Anne Moore, *Crain's Chicago Business*

"Important. . . . Egan's book serves as a reminder that the ecological universe we inhabit is vastly connected and cannot be easily mended by humility and good intentions."
　　　　　　　—Meghan O'Gieblyn, *Boston Review*

"Egan's knowledge, both deep and wide, comes through on every page, and his clear writing turns what could be confusing or tedious mate-rial into a riveting story."　　—Margaret Quamme, *Columbus Dispatch*

"A literary clarion call. . . . Egan's narrative reflects a nuanced under-standing of history and science, which is matched by his keen percep-tions about public policy."　　　　　　　—*National Book Review*

"This book feels urgent to policymakers and laypersons alike."
　　　　　　　—Kerri Arsenault, *Literary Hub*

"Alarming and powerful. . . . *The Death and Life of the Great Lakes* is an engaging, vitally important work of science journalism."
　　　　　　　—Eva Holland, *Globe and Mail*

"Marvelous. . . . Egan's book is an ecological page-turner."

—John Hildebrand, *Milwaukee Journal Sentinel*

"This is a rollicking, eye-popping, scary, sad tour of one of the world's watery wonders, the Great Lakes."

—Stephanie Hemphill, *Agate* magazine

"With narrative flair, Dan Egan tells the story of how it is that we can be so shortsighted and negligent when it comes to something as wondrous and essential as our Great Lakes and yet so industrious and inventive in trying to undo our mistakes. This is essential reading for anyone concerned about the future of our natural world."

—Alex Kotlowitz

"In this beautifully vivid portrait of the Great Lakes, Dan Egan explores one of America's most essential ecosystems, reminding us that its story—one of both harm and hope—is ultimately our own."

—Deborah Blum

"A masterpiece. Dan Egan's epic story is one of those rare books that can change the world. Rachel Carson's *Silent Spring* sparked a national revolt against toxic pesticides. Egan's work could help save the world's biggest body of fresh water. Read it if you care about this country—and our planet."

—Tim Weiner

THE
DEATH
AND LIFE
OF THE
GREAT
LAKES

Dan Egan

W. W. NORTON & COMPANY

INDEPENDENT PUBLISHERS SINCE 1923

NEW YORK   LONDON

For information about permission to reproduce selections from this book,
write to Permissions, W. W. Norton & Company, Inc.,
500 Fifth Avenue, New York, NY 10110

For information about special discounts for bulk purchases, please contact
W. W. Norton Special Sales at specialsales@wwnorton.com or 800-233-4830

Manufacturing by LSC Harrisonburg
Book design by Daniel Lagin
Production manager: Julia Druskin

Library of Congress Cataloging-in-Publication Data

Names: Egan, Dan, author.
Title: The death and life of the Great Lakes / Dan Egan.
Description: First edition. | New York, NY : W.W. Norton & Company, 2017. |
    Includes bibliographical references and index.
Identifiers: LCCN 2016039546 | ISBN 9780393246438 (hardcover)
Subjects: LCSH: Lake ecology—Great Lakes (North America) | Great Lakes
    (North America)—Environmental conditions. | Introduced organisms—Great
    Lakes (North America) | Nonindigenous aquatic pests—Great Lakes (North
    America) | Water quality—Great Lakes (North America)
Classification: LCC QH104.5.G7 E43 2017 | DDC 577.630977—dc23
LC record available at https://lccn.loc.gov/2016039546

ISBN: 978-0-393-35555-0 pbk.

W. W. Norton & Company, Inc.
500 Fifth Avenue, New York, N.Y. 10110
www.wwnorton.com

W. W. Norton & Company Ltd.
15 Carlisle Street, London W1D 3BS

9 0

The Great Lakes watershed.

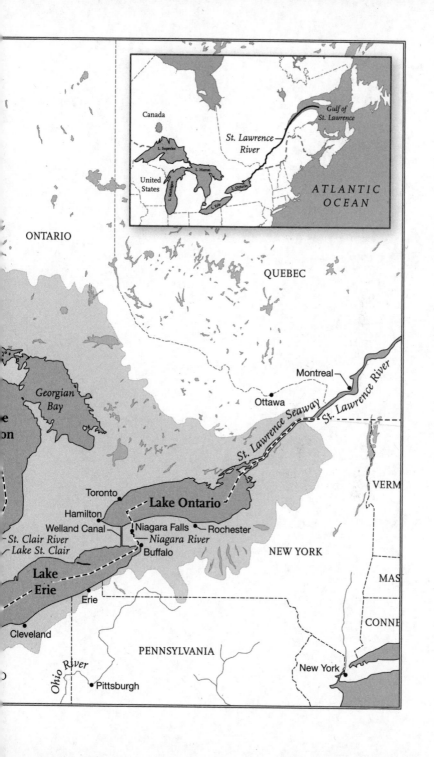

CONTENTS

INTRODUCTION   XI

PART ONE: **THE FRONT DOOR**

Chapter 1. CARVING A FOURTH SEACOAST
DREAMS OF A SEAWAY   3

Chapter 2. THREE FISH
THE STORY OF LAKE TROUT, SEA LAMPREYS AND ALEWIVES   36

Chapter 3. THE WORLD'S GREAT FISHING HOLE
THE INTRODUCTION OF COHO AND CHINOOK SALMON   75

Chapter 4. NOXIOUS CARGO
THE INVASION OF ZEBRA AND QUAGGA MUSSELS   108

PART TWO: **THE BACK DOOR**

Chapter 5. CONTINENTAL UNDIVIDE
ASIAN CARP AND CHICAGO'S BACKWARDS RIVER   151

Chapter 6. CONQUERING A CONTINENT
THE MUSSEL INFESTATION OF THE WEST   187

Chapter 7. NORTH AMERICA'S "DEAD" SEA
TOXIC ALGAE AND THE THREAT TO TOLEDO'S WATER SUPPLY   212

PART THREE: **THE FUTURE**

Chapter 8. PLUGGING THE DRAIN
THE NEVER-ENDING THREAT TO SIPHON AWAY GREAT LAKES WATER   247

Chapter 9. A SHAKY BALANCING ACT
CLIMATE CHANGE AND THE FALL AND RISE OF THE LAKES   277

Chapter 10. A GREAT LAKE REVIVAL
CHARTING A COURSE TOWARD INTEGRITY, STABILITY AND BALANCE 300

Acknowledgments   323

Notes   325

Selected Bibliography   345

Illustration Credits   349

Index   351

INTRODUCTION

There are few views that can draw noses to airplane windows like those of the Great Lakes. From on high, the five lakes that straddle the U.S. and Canadian border can appear impossibly blue, tantalizing as the Caribbean. Standing on their shores and staring out at their ocean-like horizons, it hits you that the Great Lakes are, in one significant way, superior to even the Seven Seas. The Great Lakes, after all, are so named not just for their size but for the fact that their shorelines cradle a global trove of the most coveted liquid of all—freshwater.

The world's largest freshwater system has captured the public's imagination since the first European explorers arrived on the shores of the "sweet water seas" in the early 1600s convinced—or at least everhopeful—that on their far shores lay the riches of China. In 1634 voyageur Jean Nicolet paddled his birch bark canoe across northern Lake Huron, through the Straits of Mackinac and headed for the western side of Lake Michigan—a place no white man had evidently ever set eyes upon. Nicolet arrived in a bay on the far shore of Lake Michigan apparently trying to look like a local—in a flowing Chinese robe bursting with colorful flowers and birds. Although he might have thought he had finally finished the job Columbus started a century and a half earlier, he actually landed on the southern end of an arm of Lake Michigan known as Green Bay. There is a statue today of Nicolet in that robe that stands near the reputed landing site. It's 20 minutes north of Lambeau Field, some 7,000 miles shy of Shanghai.

It's hard to fault Nicolet if he really did believe his journey had taken him to Asia, because there were no Old World analogues for the scope of the lakes he was trying to navigate. The biggest lake in France, after all, is 11 miles long and about 2 miles wide; the sailing distance between Duluth, Minnesota, on the Great Lakes' western end and Kingston, Ontario, on their eastern end is more than 1,100 miles. No, the bodies of water formally known as the Laurentian Great Lakes are not mere lakes, not in the normal sense of the word. Nobody staring across Huron, Ontario, Michigan, Erie or Superior would consider the interconnected watery expanse that sprawls across 94,000 square miles just a lake, any more than a visitor waking up in London is likely to think of himself as stranded on just an island (the United Kingdom, in fact, also happens to span some 94,000 square miles).

A normal lake sends ashore ripples and, occasionally, waves a foot or two high. A Great Lake wave can swell to a tsunami-like 25 feet. A normal lake, if things get really rough, might tip a boat. A Great Lake can swallow freighters almost three times the length of a football field; the lakes' bottoms are littered with an estimated 6,000 shipwrecks, many of which have never been found. This would never happen on a normal lake, because a normal lake is knowable. A Great Lake can hold all the mysteries of an ocean, and then some.

In 1950, when Northwest Airlines flight 2501 flying from New York City to Seattle disappeared from radio contact after it hit a summer storm over Lake Michigan, it was at the time the worst commercial aviation accident in U.S. history. The Coast Guard and Navy dispatched five ships to look for the wreckage. They dropped sonar devices, divers and drag lines into the lake to hunt for the nearly 100-foot-long fuselage that carried 58 souls.

The wreck has never been found.

Here is a different way to grasp the scale of the Great Lakes. Roughly 97 percent of the globe's water is saltwater. Of the 3 percent or so that is freshwater, most is locked up in the polar ice caps or trapped so far underground it is inaccessible. And of the sliver left over that exists as surface freshwater readily available for human use, about 20 percent of that—one out of every five gallons available on the planet—can be found in the Great Lakes. This is not an insignificant fact at a time when more than three-quarters of a billion people don't have regular access to safe drinking water.

In 1995, World Bank vice president Ismail Serageldin made a provocative prediction: "The wars of this century have been fought over oil, and the wars of the next century will be on water . . ." Perhaps. But the biggest enemy facing the Great Lakes in the early 21st century is not would-be profiteers seeking to siphon them off to make far-away deserts bloom. The biggest threat to the lakes right now is our own ignorance.

Nearly 500 years after Nicolet first nosed his canoe into the waters of Lake Michigan we are still treating the lakes the same way, as liquid highways that promise a shortcut to unimaginable fortune. Nicolet might have made an honest mistake. The same won't be said for us, because continuing to exploit the world's largest expanse of freshwater in this manner is wreaking increasingly disastrous consequences.

YOU MAY THINK YOU KNOW THE MODERN HISTORY OF THE GREAT Lakes. The story of how by the middle of the 20th century industrial and municipal pollution smothered their beaches, of how hundreds of square miles of open water at that time were so devoid of oxygen they were declared "dead," and of how the rivers that feed the lakes suffocated under chronic slicks of chemicals and oils prone to combust. And then the story of their revival, of how all the industrial plundering and wanton polluting finally spurred passage of the landmark Clean Water Act of 1972.

That law did indeed bring dramatic reductions in the wastes tumbling into the lakes, and their recovery was as fast as it was dramatic. This is why lakefronts from Toronto to Milwaukee today glimmer with the glass of luxury condos and office towers, and why the land they sit atop is among the most expensive real estate in the Midwest. This is why Cleveland's Cuyahoga River, which famously exploded in flames, now draws more fishing lines than punch lines. And this is why when you cruise along Chicago's Lakeshore Drive on a hot summer afternoon you will see hundreds of people lounging on the beach and splashing in the Lake Michigan surf, all literally in the shadows of the John Hancock Center and its neighboring skyscrapers. It all gives the impression that humans and the lakes have finally learned to get along. It's a mirage.

The story of *The Death and Life of the Great Lakes* takes you beneath the lakes' shimmering surface and illuminates an ongoing and unparalleled ecological unraveling of what is arguably North America's most precious natural resource. It's about how the Great Lakes were resuscitated after a century's worth of industrial abuse only to be hit with an even more vexing environmental catastrophe.

Tragically for the Great Lakes, the Clean Water Act helped to lull much of the public into thinking that the lakes had hit their nadir and were on their way to recovery throughout the 1970s, '80s and '90s. But the law—or, more specifically, the agency charged with enforcing it—in fact did unfathomable damage to the lakes. It turned out that federal environmental regulators decided to exempt one industry's form of "living pollution"—biologically contaminated water discharged from freighters. This exemption included overseas ships sailing up the man-made St. Lawrence Seaway that links the lakes to the Atlantic Ocean and to ports around the globe—China, finally, included.

The Seaway, which opened in 1959 with all the hoopla of a lunar landing, never lived up to its hype. Today an average of only about two overseas freighters visits the Great Lakes each day during the nine-month shipping season (the Seaway closes each winter due to ice). The oceangoing ships that make the trip into the lakes aren't super-sized container vessels hauling high-value goods like Sonys and Toyotas. The cramped shipping channels of the Seaway can only accommodate 1930s-sized freighters, and these ships typically bring in foreign steel and haul out U.S. and Canadian grain. But the Seaway ships have also been hauling something not listed on their manifests—noxious species from ports all over the world that are inexorably unstitching a delicate ecological web more than 10,000 years in the making.

The Environmental Protection Agency apparently received no statutory authority from Congress to make this exemption for Seaway vessels and other freighters visiting U.S. waters. The agency, for whatever reason, decided to quietly tweak the regulations the year after the Clean Water Act was passed, probably to save some hassle and money by giving the shipping industry a free pass to dump their ship-steadying ballast water. The rationale at the time was that ballast tanks weren't full of noxious stuff like oils or acids. They held nothing but seawater. The folly here is that ballast water isn't just water. It swarms with perhaps the most potent pollutant there is: DNA.

It would be hard to design a better invasive species delivery system than the Great Lakes overseas freighter. The vessels pick up ballast water at a foreign port to balance less-than-full cargo loads. When the ships arrive in the Great Lakes, cargo is taken onboard and the ballast water—up to 10 Olympic swimming pools' worth per ship—and all the life lurking in it gets set loose in the lakes. As one exasperated Great Lakes biologist once told me: "These ships are like syringes."

The Great Lakes are now home to 186 nonnative species. None has been more devastating than the Junior Mint–sized zebra and quagga mussels, two closely related mollusks native to the Black and Caspian Seas. A college kid on a field trip in the late 1980s was the first to discover them in the Great Lakes. In less than 20 years the mussels went from novel find to the lakes' dominant species. Sandy beaches still rim the lakes, but if Lake Michigan, for example, were drained it would now be possible to walk almost the entire 100 miles between Wisconsin and Michigan on a bed of trillions upon trillions of filter-feeding quagga mussels.

The mussels, which have no worthy natural predators in North America, have transformed the lakes into some of the clearest freshwater on the planet. But this is not the sign of a healthy lake; it's the sign of a lake having the life sucked out of it.

The cumulative toll of the EPA's long-standing ballast water exemption (an exemption that it is now under court order to remedy—someday, or decade) is far more dire than a burning chemical slick on a polluted river. Native fish populations have been decimated. Bird-killing botulism outbreaks plague lakeshores. Poisonous algae slicks capable of shutting down public water supplies have become a routine summertime threat. A virus that causes deadly hemorrhaging in dozens of species of fish, dubbed by some scientists the "fish Ebola," has become endemic in the lakes—and threatens to spread across the continent.

ICONIC DISASTERS HAVE A HISTORY OF PROMPTING GOVERNMENT action. Three years after the Cuyahoga River fire of 1969, Congress passed the Clean Water Act. Two decades later, when the *Exxon Valdez* ran aground and dumped 10.8 million gallons of crude oil into Alaska's

Prince William Sound, images of cleanup crews using paper towels to cleanse tarred birds helped press Congress into doing something it should have done years earlier. It mandated double-hulled oil tankers.

But the disaster unfolding today on the Great Lakes didn't ignite like a polluted river or gush like oil from a cracked hull, and so far there is no galvanizing image of this slow-motion catastrophe, though a few come to my mind. One is the bow of an overseas ship easing its way into the first navigation lock on the St. Lawrence Seaway, the Great Lakes' "front door" to fresh waves of biological pollution. Another is a satellite photo of a green-as-paint toxic algae slick smothering as much as 2,000 square miles of Lake Erie.

Yet another is the grotesque mug of an Asian carp, a monster-sized carp imported to the United States in the 1960s and used in government experiments to gobble up excrement in Arkansas sewage lagoons. The fish, which can grow to 70 pounds and eat up to 20 percent of their weight in plankton per day, escaped into the Mississippi River basin decades ago and have been migrating north ever since. They are now mustering at the Great Lakes' "back door"—the Chicago canal system that created a manmade connection between the previously isolated Great Lakes and the Mississippi basin, which covers about 40 percent of the continental United States. The only thing blocking the fish's swim through downtown Chicago and into Lake Michigan is an electrical barrier in the canal—one that has a history of unexpected shutdowns.

The Chicago canal has also turned the Great Lakes' ballast water problem into a national one, because there are dozens of invasive species poised to ride its waters out of the lakes and into rivers and water bodies throughout the heart of the continent. Species like the spiny water flea, the threespine stickleback, the bloody red shrimp and the fishhook water flea. All organisms you probably haven't heard of. *Yet.*

Few out West, after all, had ever heard of quagga mussels—until they tumbled down the Chicago canal and metastasized across the

Mississippi basin and, eventually, into the arid West, likely as hitch-hikers aboard recreational boats towed over the Rocky Mountains. The mussels have since unleashed havoc on hydroelectric dams, drinking water systems and irrigation networks in Utah, Nevada and California and the federal government estimates that if the mussels make their way into the Northwest's Columbia River hydroelectric dam system they could do a half billion dollars of damage—per year.

The engineers, water managers and biologists out West who view the Great Lakes as a beachhead for invasions that inevitably go national look at the Seaway and are incredulous at the recklessness of leaving this door to the entire continent open. So are most people, once they understand the vastness of the problem, and the tiny industry causing so much of it.

If we can close these doors to future invasions, we may give the lakes, and the rest of the country, time to reach a new equilibrium, a balance between what is left of the natural inhabitants and all the newcomers (there are already signs in some areas of the lakes that native fish are adapting to a diet based on zebra and quagga mussels). And if we do this, then we can focus on the major problems that still plague the lakes, which include the overapplication of farm fertilizer that is helping to trigger the massive toxic algae outbreaks, the impact a warming globe is having on the lake's increasingly unstable water levels and the need to protect lake waters from outsiders seeking to drain them for their own profit.

Like generations of the past, we know the damage we are doing to the lakes, and we know how to begin to stop it; unlike generations of the past, we aren't doing it.

This situation reminds me of those black-and-white photos of settlers standing next to a mountain of bison skulls during the Great Plains slaughters of the late 1800s. The skulls were considered garbage at the time. Some were crushed and used as a cheap form of pavement, before they became so rare so quickly that by the early 1900s they were already fetching $400 apiece from collectors trying to cling to a fragment of what had been squandered.

Every time I see one of those pictures I'm struck with two thoughts. *What in the hell were they thinking?* And, more importantly: *Is what we are doing to the Great Lakes today going to leave our own great-grandchildren equally baffled?*

PART ONE

THE FRONT DOOR

Chapter 1

CARVING A FOURTH SEACOAST
DREAMS OF A SEAWAY

I n 1957 legendary CBS newsman Walter Cronkite—lauded as *the most trusted man in America*—stared into the camera and told viewers that the "greatest engineering feat of our time" was under way. He wasn't talking about the Soviet Union rocketing the stray dog Laika into orbit, or that year's development of the first wearable pacemaker, or the recent opening of the United States' first commercial atomic power plant. He was talking about humans "conquering" nature on a scale and in a fashion never before attempted.

"Right now the greatest concentration of heavy machinery ever assembled—over 3,000 pieces of equipment—are at work on one of the greatest projects in the history of mankind," Cronkite said as he stood in front of a map of the deep blue Great Lakes and the even deeper blue Atlantic Ocean. He fixed his eyes on the camera and spoke boldly of a construction project that would, in effect, do no less than move the Atlantic Ocean more than 1,000 miles inland, to the middle of North America.

The idea was to scrape and blast a navigation channel along and through the shallow, tumbling St. Lawrence River that flows from the

Great Lakes out to the ocean in a manner that would allow giant freighters to steam from the East Coast into the five massive freshwater inland seas. This manmade nautical expressway, as narrow as 80 feet in places and, in one particularly tight section, crossing *over* a roadway, would open up some 8,000 miles of U.S. and Canadian coastline to ships from around the world. The hope was that essentially landlocked Great Lakes cities like Chicago, Cleveland, Detroit and Toronto would blossom into global ports to rival commercial hubs such as New York, Rotterdam and Tokyo.

The project, Cronkite told his viewers, was big, big as "reshaping a continent, completing the job nature had begun thousands of years ago—of creating an eighth sea . . . a sea of opportunity!"

More than a half century later, the hoped-for flood of global cargo has yet to roar into the lakes from overseas, but something else has—an environmental scourge whose scope and costs are spreading by the day. The St. Lawrence Seaway, you see, didn't conquer nature at all.

It unleashed it in the form of an ecological catastrophe unlike any this continent has seen.

IT IS HARD TO FAULT CRONKITE TODAY FOR HIS OPTIMISM, BECAUSE the nautical magic he and so many others were convinced the Seaway would uncork had happened before. Some six million years ago, the Mediterranean Sea itself was isolated from the Atlantic Ocean. It was little more than a salty puddle at the bottom of a vast basin laced with dusty canyons, some of which plunged more than a mile below sea level. This arid wasteland had previously been a massive Atlantic Ocean inlet, as it is today. But then a tectonic fusion of Africa and Europe created a narrow strip of land that plugged the Mediterranean's connection to the Atlantic Ocean near what is now the Strait of Gibraltar. This pretty much killed the ancient Mediterranean Sea, which owed its existence to a constant inflow of ocean water, just as it does today. With that Atlantic input plugged, the rivers feeding the suddenly land-locked basin proved too feeble to keep pace with evaporation, and the sea all but vanished in about 1,000 years—which is to say, geologically speaking, nothing. But on a human scale the sea would have shrunk at an imperceptibly slow pace; each day on its shores would have seemed exactly like the last.

The Mediterranean Sea basin, one popular theory goes, remained in this desiccated state for the next 700,000 years or so. But about 5.3 million years ago a seismic hiccup at the Gibraltar isthmus opened a small channel for the Atlantic Ocean to begin dribbling back in. The trickle soon turned to a torrent, many of today's geologists reckon, as an ever-widening and deepening tongue of saltwater roared back into the basin with incomprehensible speed, volume and violence. It carried the equivalent of some 40,000 Niagara Falls flowing at about 90 miles per hour. This all happened around the time our ancestors' thigh bones formed a bridge with their hips strong enough to allow them to walk upright and, perhaps—if any of them happened to be in the area at the time the Atlantic came roaring back—to run.

At the peak of the Atlantic cascade the new Mediterranean Sea was rising at a rate of about 30 feet per day, and geologists hypothesize that the entire basin—roughly 2,500 miles long and 500 miles wide—could have filled to sea level in less than three years.

The Mediterranean's revival indubitably wrought devastation for the terrestrial creatures scratching out a life in the scorched basin, including dwarf elephants and hippos. But it proved a boon for the dolphins and fish and even microscopic life sucked in from the North Atlantic. The devastation also, eventually, opened the door for civiliza-tion to blossom, because the Mediterranean Sea connected cultures and economies in a manner that would not have been possible had the basin remained a desert. Today the Mediterranean gives 21 countries from three continents nautical access to each other and—thanks to the eight-mile-wide Strait of Gibraltar carved by the Atlantic Ocean—to the rest of the globe.

About 7,600 years ago, the Black Sea was isolated from the Atlantic Ocean. It was an inland freshwater lake cut off from the Mediterranean Sea to the west by a spit of land called the Bosporus Valley. At the peak of the last ice age some 20,000 years ago, so much of the earth's water was tied up in glaciers that, according to some estimates, sea level was nearly 400 feet lower than it is today. As the glaciers melted and the oceans rose, so did the Mediterranean. And eventually the Mediterranean did to the Black Sea what the Atlantic Ocean had done to it more than 5 million years earlier: it came crashing in.

The speed with which this happened, as well as its scale, is a matter of some controversy, but a popular hypothesis is that the salty water tumbled in at a force equivalent to 200 Niagara Falls. The inundation that submerged some 60,000 square miles under hundreds of feet of water happened so swiftly—some geologists estimate the sea was rising at a rate of about six inches per day—that it would have sent scrambling any humans who had found the lakeshore an oasis in an otherwise parched landscape. The salty water also ravaged the lake's freshwater biological community, rendering extinct the species that could not adapt and sending others—like the Black Sea sturgeon—darting for safety in the freshwater rivers that still feed the sea today.

To call this a natural disaster of biblical proportions is what two Columbia University geophysicists did when they published a book in 1998 titled *Noah's Flood*. They argue that this geologic event, which is commonly known as the Black Sea Deluge, could be the inspiration for the great flood stories of the past, including the one in the Book of Genesis. That two geologists contend a real flood could be tied to a story in the Bible was not without some controversy in the academic community—and, of course, among believers. But leaving aside any biblical implications, their geological evidence for the disaster itself is solid. And, like the torrent that roared through the Strait of Gibraltar millions of years earlier, there was an upside to it; the merging of the Black and Mediterranean Seas opened up a critical nautical link stretching from Asia to the Atlantic Ocean. Today the Bosporus Strait is one of the world's busiest shipping channels, with freighters sailing from the once-landlocked Black Sea to ports around the globe.

About 200 years ago, North America's Great Lakes, the largest expanse of freshwater in the world, remained essentially isolated from the Atlantic Ocean. For thousands of years, the five inland seas wrapped by more than 10,000 miles of shoreline (islands included) sat cloistered in the middle of the continent. The four "upper" lakes—Erie, Huron, Michigan and Superior—lie some 600 feet above the level of the ocean, which made them unreachable from the Atlantic by boat. Much of that elevation is gained at the dolomite cliffs that are Niagara Falls, over which the collective outflows of all those lakes tumble on their way into Lake Ontario and from there down the thundering St. Lawrence River on their rush to the ocean.

Like the plugs of land that once isolated the basins that are now the Mediterranean and Black Seas, erosion has been having its way with Niagara Falls. It is expected the falls will disappear in about 50,000 years—which is to say, geologically speaking, pretty soon. When that happens, the cliffs that have for millennia separated the upper Great Lakes from the Eastern Seaboard will be gone. All that will remain is a fast-flowing, ever-eroding riverbed that will draw the lakes, every day, one step closer to sea level. How this all precisely plays out in terms of perhaps opening a nature-carved sailing route between the middle of the continent and the ocean is a matter of geological conjecture that won't be answered for eons—an unbearably long period for the 19th- and 20th-century Great Lakes politicians and businessmen who were not content to leave the lakes as they had found them, as isolated inland seas upon which giant cargo boats could float from one Midwestern city to another, but never out to the ocean.

Their idea was to finish the job nature started when the last glaciers carved out the Great Lakes basins 10,000 years ago. Their dream was to create, by the hand of man, a North American "Fourth Seacoast," thus flexing the Midwest's burgeoning manufacturing might across the globe, prying open new markets in far-away cities and squeezing from them all manner of exotic bounty. They lusted for their own Mediterranean, for their own Strait of Gibraltar or Bosporus to emerge, but they were not willing to wait for such a natural disaster to unfold.

So they hatched an unnatural one.

THE MAP PRACTICALLY TAUNTED THE UNITED STATES AND CAN-ada to build the St. Lawrence Seaway. The tendril of blue reaching out to the Atlantic Ocean from Lake Ontario—the Gulf of St. Lawrence and the St. Lawrence River that feeds it—stretches some 1200 miles inland. And, on a map, that flat ribbon of blue continues on from Lake Ontario, through Lake Erie, into Lakes Michigan and Huron and all the way across Lake Superior to Duluth, Minnesota, on its western shore. If you were to plot a voyage based on this map, you might assume you could paddle or sail your way from the Atlantic Coast almost to the dead-center of North America—a distance of about 2,300 miles. And, in a boat, you would indeed find waters as flat as those on a map for almost half the trip. But everything changes about 1,000 miles inland.

Jacques Cartier, the first European known to reach the area by boat, learned this firsthand when the yawning river up which he sailed so effortlessly in 1535 turned narrow and vicious in an instant. The 44-year-old lifelong explorer, descended from a long line of mariners, had been handpicked by France's King Francis I to find a nautical shortcut across North America to tap the riches of Asia and, of course, to pick up any gold and silver nuggets he found along the way.

The summer before, Cartier led a two-boat expedition across the Atlantic Ocean that probed as far west as the Gulf of St. Lawrence but stopped short of sailing up the St. Lawrence River that feeds it. He returned to France that fall, his cargo hold empty of precious metals but his head filled with Native American tales that a vast sea did indeed lie at the head of the St. Lawrence River. The next year the king gave Cartier 110 men and three boats, including one specially modified to sail up rivers.

The boat wasn't special enough for the St. Lawrence job. No boat would be for hundreds of years.

Just upstream from the island that is now downtown Montreal, Cartier encountered a set of oversized rapids, a word that doesn't adequately capture how angry and impenetrable to upstream navigation this river was. There were waves approaching six feet in height, like those you'd expect to see on an ocean beach when the red no-swim flags are snapping. But these waves didn't crash. They forever arced, never tumbling into a froth that might be breached by some well-timed paddling. It was a standing, ever-rolling wall of water created by the plunging St. Lawrence riverbed. Cartier remained convinced there were loads of gold beyond the waves and, perhaps, the fabled shortcut to Asia, but the water was so violent it stopped him mid-voyage. He turned around and sailed back down the river. The French explorers who came after remained convinced that somewhere beyond this violent water lay the riches of China, and the rapids today remain named Lachine, which is French for that promised land.

The voyageurs who eventually pressed further inland by portaging their birch bark canoes around the rapids quickly learned that far upstream lay something almost as miraculous: a set of connected fish-filled freshwater seas larger than any explorer had ever encountered, surrounded by forests of pine and hardwoods that teemed with game—and pelts—on a scale incomprehensible in Europe. But the Lachine Rapids were just the first line of defense for what would one day be called the Great Lakes. In the thousand miles or so it took to sail from the Atlantic to Montreal, the St. Lawrence River rose all of 18 feet. In the 189 miles upstream from Montreal to Lake Ontario the river climbed some 245 feet in a series of impassable torrents.

Then the real whitewater started. On the far side of Lake Ontario lay another frothing river that gained about 160 feet in just 35 miles. Anyone who tried to paddle or portage up that gorge hit a wall. Literally.

Niagara Falls are what made the Great Lakes unique in the natural world. The falls are the most famous 1,100 yards of a 650-mile-long ridge of sedimentary rock arcing from western New York, into the province of Ontario, and down into Wisconsin. This escarpment is the rim of a 400-million-year-old seabed that cradled a shallow, tropical ocean that once sloshed across what is today the middle of North America. At about 170 feet high, the falls that tumble over the Niagara escarpment near present-day Buffalo, New York, are nowhere near the world's tallest or even largest by volume. But they were among the most ecologically important because they created an impassable barrier for fish and other aquatic life trying to migrate upstream from Lake Ontario into the other four Great Lakes.

Other giant freshwater bodies that have evolved over tens of thousands or even millions of years have been subjected to epic changes in temperature, salinity, water levels as well as wave upon wave of invading and evolving organisms, all in a manner that leaves those water bodies inhabited by a cast of species steeled by the crucible of evolution. This gives them something of an "immune system" when it comes to maintaining ecological stability in the face of disruptions from the outside world. The Great Lakes of Cartier's time, on the other hand, were what biologists today call "ecologically naïve." This means the lakes were inhabited by fish and other aquatic species whose isolation left them uniquely exposed to foreign perturbances. None of this, of course, was pondered by the early explorers desperate to exploit their ecological bounty.

The ditch-digging to open a commercial passage into the Great Lakes by first building a canal around Lachine Rapids started in 1689 but was scuttled soon after when French crews equipped with only the crudest of tools ran into more stubborn rock than expected—and attacks from Native Americans. Work on that tiny section of river alone would sputter all the way into the 1800s, even as progress was made in taming other St. Lawrence rapids farther upstream toward Lake Ontario, particularly after the English captured Canada from the French in 1763.

In the next two decades the English military, eager to maintain control of the region in the face of rebellion from the 13 U.S. colonies, began chewing its way upriver to supply troop outposts. The first big bite through the St. Lawrence barrier came in 1781, during the height of the Revolutionary War, with the opening of a canal running parallel to the northern bank of the St. Lawrence River, about 25 miles upstream from Montreal. It stretched scarcely the length of a football field and was less than six feet wide and three feet deep. But it was not the size of this little detour around the rapids that made the canal so significant. It was the technology built into it. It had three navigation locks that may well have been the first constructed on this continent.

In a navigation lock, an upriver-bound boat enters a watertight chamber that has a downstream front door and an upstream back door. At the time an upriver boat noses through the open downstream door and into the chamber, the upstream door is already closed. Once the boat is fully within the chamber the downstream door is closed as well. Then a gate is opened to a sluice fed by river water on the upstream side and the chamber is filled until it matches the water level on the upper side. The upstream doors swing open so the boat can smoothly progress upriver. Downstream boats go through the process in reverse. The only engine a system like this needed was gravity to send the water into and out of the chambers, and human muscle to crank the lock doors open and shut.

This first short canal allowed a boat to ascend, or descend, a mere six feet before it returned to the main river channel. It was a modest breach in the defense of the Great Lakes, but the canal building inexorably progressed upriver and soon stretches that had been accessible only by birch bark canoes that could be portaged around rapids were being plied by flat-bottomed rowboats 40 feet long. These "bateaux" had a draft of less than three feet but each could carry more than three tons of cargo—furs and timber downstream and food, tools and people upstream. By 1800, the river beyond Montreal had become accessible to larger Durham boats (the kind George Washington used in 1776 to cross the Delaware River in his Christmas night raid) that could be equipped with a sail and haul more than double the cargo of a bateau. Yet at the beginning of the 19th century the Lachine Rapids at Montreal had yet to be breached with an adequate canal, and in other particularly rough stretches along the St. Lawrence River cargoes had to be unloaded as the boats were tugged through the whitewater. It took about 12 days to make the 180-mile trip that started just above Lachine to Lake Ontario.

Moving cargo and people along the river got much easier in 1825, when the Lachine Rapids were finally bypassed with their own lock and canal system. The manmade waterway was more than 8 miles long and included seven lock chambers that collectively raised boats about 45 feet. Completion of the canal finally provided boats a reliable float from the Atlantic Ocean into Lake Ontario, and the impact this had on goods flowing into North America's interior was almost immediate. By the early 1830s about 2,000 trips on the river between Montreal and Lake Ontario occurred annually and 24,000 tons of cargo was hauled—four times the volume of traffic in the year before the Lachine canal opened. It took a century and a half of chipping rock and plowing earth to put this crack in the geographic barrier protecting the Great Lakes from the outside world below, but it was about to turn into a chasm.

THEY MIGHT BE CALLED THE GREAT LAKES, BUT THE FIVE INLAND seas are essentially one giant, slow-motion river flowing west-to-east, with each lake dumping like a bucket into the next until all the water is gathered in the St. Lawrence River and tumbles seaward.

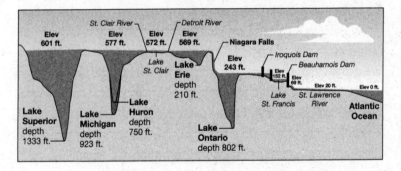

The surface elevation (in relation to sea level) of the Great Lakes, and the natural barrier at Niagara Falls.

Lake Superior sits at the system's headwaters. It is about 350 miles long and 160 miles wide, and it holds enough water to submerge a landmass about the size of North and South America under a foot of water. The lake basin might have been carved by the glaciers, but the 1,300-foot-deep sea is not simply an oversized puddle of ancient ice melt. Lake Superior is a dynamic system, ever filling up with precipitation and stream inflows, and ever flowing out toward the Atlantic.

Lake Superior inflows are balanced by its outflows down the St.

Marys River. Along its 60-mile course the river drops about 22 feet in elevation until it spills into Lake Huron, which is, really, the same body of water as Lake Michigan. They are two lobes of the same big lake connected at the five-mile-wide Straits of Mackinac. Both Michigan and Huron flow into the St. Clair River, which flows toward Lake Erie, whose elevation is only about 9 feet lower than that of Michigan and Huron. All of Erie's waters move eastward toward its outlet—the Niagara River that plunges 325 feet into Lake Ontario. Most of that drop happens midway down the river at Niagara Falls.

For thousands of years there was no way anything in or on the water below the falls could breach this barrier between Lake Ontario and the upper Great Lakes, but its collapse came swiftly, and it came on the United States' side of the border.

President George Washington was among the first to grasp the danger of allowing settlement of American territories west of the Appalachian Mountains to take its own course. Washington believed there was no reason the inland immigrants on that isolated frontier, severed from the 13 seaboard states by the mountain crests of the Appalachians, would maintain allegiance to their new country instead of the settlers allied with Great Britain to the north, or with the Spanish to the south. He wanted a canal extending west from the Mid-Atlantic's Potomac River, but he recognized that a connection to the West had to be made, one way or the other—and in one place or another.

"I need not remark to you Sir, that the flanks and rear of the United States are possessed by other powers, and formidable ones too; nor how necessary it is to apply the cement of interest, to bind all parts of the Union together by indissoluble bonds, especially that part of it, which lies immediately west of us . . ." Washington wrote to Virginia Governor Benjamin Harrison in the fall of 1784. "The Western settlers, (I speak now from my own observation) stand as it were upon a pivot; the touch of a feather, would turn them any way."

Black powder and pickaxes affixed these western settlements to the United States. It took 40 years and it did not follow the route Washington championed, but his dream of an umbilical cord stretching westward from the colonies to the interior was realized with the opening of the Erie Canal in 1825. Much of the New York state route between Lake Erie and the Atlantic Coast had already been carved by nature. Like the lower St. Lawrence River, the Hudson River rolls ever so gently into the sea, its tilt so tame that ocean tides push upriver as far as Albany. That made for 145 miles of smooth sailing into what was then the deep American interior. Due west and through some 300 miles of thick forest and stubborn Allegheny Mountains, lay the outpost of Buffalo on the shore of Lake Erie. The overland trip by stage coach between Albany and Buffalo took about two weeks in the early 1820s, most of it over roads so rough that passengers often had to get out and push the carriage up a bumpy slope, through mud and over ruts. There had to be a better way.

New York governor and one-time mayor of New York City DeWitt Clinton gets much of the credit for spearheading construction of the state-funded Erie Canal across this rough route, and he was the politician who sold the concept to the public. But the engineering idea that made it possible was hatched from a prison cell. Jesse Hawley, a flour merchant in western New York, had gone broke trying to move his product down the mess of roads and trails that wended their way out of the wilderness of western New York. Hawley spent 20 months in debtors' prison beginning in 1807, and while there he scratched out more than a dozen letters to the *Genesee Messenger* arguing for construction of a canal linking the Hudson River to the Great Lakes. He wrote that he was motivated by wanting to atone for having led a life of "little purpose" up to that point. The letters laid out the general route that the Erie Canal would eventually take. Hawley knew he was thinking big, acknowledging later in life that his argument was initially received as "the effusions of a maniac." But there was a genius in it. The way he saw it, God put the Great Lakes so high above sea level for one reason—to provide the energy to fill the locks to lift the boats. Had Lake Erie been at an equal level in elevation to the Hudson River but still separated by a mountain range, such a canal would not have been possible. But once men who knew how to build navigation locks went to work, the upper lakes' greatest line of defense to the outside aquatic world proved to be their greatest weakness.

"It appears the Author of nature, in forming Lake Erie with its large head of waters into a reservoir," Hawley wrote, ". . . had in prospect a large and valuable canal, connecting the Atlantic and the continental seas, to be completed at some period in the history of man, by his ingenuity and industry!"

The idea was derided across the nation as impractical, if not technologically impossible. But it stirred the passions of the only man who mattered—Clinton. As mayor of New York in the early 1800s, the young lawyer initially saw the canal as a means for his city to keep pace with Boston and Philadelphia. But by 1816 he had sold the canal as essential to the economic future of the nation and had won financial backing for it from Congress, though that support was snuffed by a veto from President James Madison.

Clinton, who became New York governor in 1817, pushed forward with the canal as a state project that began that year on July 4. Public support for an enterprise the press mocked as "Clinton's Folly" would wane in the following years to the point that Clinton lost his office. But as his vision took shape in the form of a 40-foot-wide ditch wending hundreds of miles through the western New York wilderness, enthusiasm for the canal—and its deposed champion—soared. Clinton won reelection as New York governor in 1825—eight years after canal construction started and just in time for its opening ceremonies in Buffalo. On October 26, 1825, at precisely 10 a.m., the first gates on the 83-lock system swung open and Lake Erie water entered the canal. Clinton and his entourage climbed aboard the *Seneca Chief*, a barge tugged by four gray horses, and headed for Albany at a speed of about four miles per hour.

Their departure was marked by a cannon blast, followed by another farther downstream once that first boom was heard, and so on, all the way down the canal's path to Albany, and then down the Hudson River to New York Harbor. It took about 90 minutes for the chain of cannon reports to hit New York City, which responded with its own blast that started a reverse, upstream-bound string of booms. Buffalo and New York City—the East Coast and the Western frontier—were now linked by a water road smooth as any modern interstate.

When the party got to New York City 10 days later, Clinton hoisted a green cask containing water drawn from Lake Erie. He tipped it into the sea. "This solemnity, at this place, on the first arrival of vessels from Lake Erie," he proclaimed as he splashed the Great Lakes water into the harbor, "is intended to indicate and commemorate the navigable communication which has been accomplished between our Mediterranean seas and the Atlantic Ocean."

It was just a dribble, but it was also a watershed event not unlike what had happened some five million years before on the other side of the ocean when the first drops of Atlantic waters crested the divide separating it from the dried-up Mediterranean basin. Less well documented, but far more portentous, is what happened on the *Seneca Chief*'s return trip. A judge from Buffalo brought back a cask inscribed with the words "Neptune's Return to Pan"—referring to the respective mythological gods of the sea and of the woods—filled with water from the Atlantic Ocean. The *Seneca Chief* reached Buffalo on Wednesday, November 23. Two days later the barge, loaded with dignitaries and pulled by a fleet of sailboats, pressed on into the open waters of Lake Erie. The judge made mention of the mixing of the waters in New York

City three weeks earlier, and then proclaimed: "We, in return, now unite those of the ocean with the Lake.

"This, fellow-citizens, closes the ceremonies which have grown out of an event hereafter to be held in grateful remembrance, and commemorated by annual demonstrations of gratitude, as one of the most important which has distinguished the history of mankind, and one from which not only the present, but generations yet unborn, even to the latest posterity, are to derive innumerable blessings."

And—it would come to pass—incalculable curses.

THE ERIE CANAL STRETCHED 363 MILES INLAND FROM ALBANY to Buffalo, climbed 568 feet in elevation and was 40 feet wide and a mere 4 feet deep, but it is hard to overstate the impact this trickle out of the continent's interior had on the United States. Some 40,000 people sailed on the Erie Canal in its first year. It slashed a bumpy two-week ride from Albany to Lake Erie to a five-day glide. But the canal wasn't just about squeezing time from the trip; it was about expanding the volume of goods moving between the deep interior and the coast. A single Erie barge could carry 30 tons, dropping the price to move a ton of freight from Buffalo to New York from about $100 to $10. In its first year alone there were about 7,000 boats operating on a canal that was so instantly successful in drawing business that within a decade tolls covered its $7 million construction cost. By 1845 more than 1 million tons moved on the canal annually, and that figure reached 2 million tons just seven years later.

Just as along a well-traveled highway, towns thrived along the canal. Think of a major city in the state of New York and it likely sprouted along the waters it connected—Rochester, Syracuse, Utica, Buffalo, Albany, and, of course, New York City. The canal had an equally big impact on the Great Lakes themselves. Once a connection between the coast and Lake Erie was secured, goods and people could float from New York Harbor all the way to Detroit, Chicago and Milwaukee, because the nine-foot rise in the river system between Lake Erie and Lakes Huron and Michigan was naturally navigable.

Canada did not sit idle as tens of thousands of Americans started to flood into the continent's interior, and millions of tons of grain, furs and forest flowed out. In 1824, less than a year before the ceremony celebrating the marriage of the waters in Buffalo, the Canadians went to work digging their own canal into Lake Erie, one that would cut across a hilly, narrow spit of land between Lakes Erie and Ontario. The Welland Canal was as much a hydraulic elevator as it was a canal. It was a system of 40 locks built specifically to bypass Niagara Falls and hoist mammoth boats 325 feet up the rock ledge separating the two smallest Great Lakes. The Welland's locks were far bigger—110 feet long and 8 feet deep—than those on the Erie Canal. This is because the Welland was built for giant freight-carrying schooners of the time; the Erie Canal existed to ferry goods down a tiny, tame channel on specially built, comparatively small barges. In this sense, the Welland, which opened in 1829, was a much more ambitious project than the Erie Canal. The idea behind it was not to just link two Great Lakes. The Welland, coupled with lock expansions downstream on the St. Lawrence River, was designed to provide giant sailing vessels and, soon enough, steamers, a direct connection between the Great Lakes and the Eastern Seaboard—and beyond.

This was both a promise and a problem that would chronically haunt the Welland Canal—all the way up until today. No matter how large the locks and canals grew in the Canadians' St. Lawrence shipping corridor, they were always doomed to become too small as the size of the world's cargo-carrying fleet inexorably grew.

By 1850 Canada's Welland Canal and St. Lawrence River locks were large enough to handle ships nearly 150 feet long and 26 feet wide, and by the early 1860s schooners were commonly sailing from the Great Lakes to Europe, hauling abroad things like beef, salt, lumber and grain and returning with steel and textiles. Promising as that early overseas traffic was, the number of such trips was destined to shrink as the world's fleet expanded. Even ships built to sail only between ports within the Great Lakes soon grew too wide or deep to squeeze through the Welland Canal, which was rebuilt once again in 1887 to lock dimensions of 270 feet long and 14 feet deep.

On the U.S. side of the border, the Erie Canal was expanded in 1862 so its locks were 70 feet wide and about 7 feet deep. That increased the cargo-carrying capacity of Erie barges from 30 tons in 1825 to 240 tons. The United States went forward in 1903 with yet another Erie expansion that was completed in 1918. The new canal could handle barges carrying 3,000 tons—100 times the size of the vessels on the original Erie Canal. Despite the upgrades, the new canal would be made obsolete later in the 20th century by trains and roads that could move goods much more quickly, and—equally importantly—do it throughout the winter months when the canal froze solid.

While the Canadians' Welland Canal and St. Lawrence locks were plagued by similar winter shutdowns, the Canadians pressed on with more expansions. Construction on a fourth Welland Canal began in 1913 and lasted until 1932. The first boat that nosed into its locks was 633 feet long and 70 feet wide, and it drew about 19 feet of water. It carried about 15,000 tons of wheat.

The problem was that this boat was basically an oversized ship in a bottle; it could roam across all five Great Lakes but couldn't squeeze through the old locks along the St. Lawrence River. This was a plug navigation advocates on both sides of the river that doubled as the U.S.–Canadian border were just itching to pull. The idea was to, once and for always, create a "Seaway" deep and wide enough to give the largest freighters of the day unfettered access from the Atlantic Coast to the heart of the continent.

"Nature has already done most of the work of building that seaway," Hanford MacNider, the former U.S. ambassador to Canada, proclaimed in 1939. "Let's finish the job!"

But the idea of a North American Mediterranean didn't sit well with U.S. politicians on the East Coast who feared it would compete with their own port cities, and throughout the first half of the 20th century Congress repeatedly turned down overtures from Canada to work together to expand the St. Lawrence locks and channels. After yet another Congressional rejection in the summer of 1952, Ontario Premier Leslie Frost had enough. "Our good neighbors to the south have decided, in their wisdom, not to come in with us," Frost steamed to the Canadian Broadcasting Company in June 1952. "They have made that decision. Now we ask them to please get out of the way and let us get on with the job."

Newly elected President Dwight D. Eisenhower had no intention of letting the Canadians dig a navigation corridor along the international border that would allow foreign vessels to sail within yards of U.S. soil and provide global access to the shared Great Lakes. "If Canada proceeds unilaterally, the United States would be precluded from exercising an equal voice in the control of traffic through the Seaway, not only in time of peace, but also when the United States is at war," warned Eisenhower's National Security Council planning board in April 1953. The Eisenhower Administration also worried that if iron ore deposits in Minnesota and Michigan's Upper Peninsula dried up, that could cripple the Midwest steel industry, which, in the context of the Cold War, the president's advisors viewed as "the most strategic of all strategic industries." Eastern Canada, however, had bountiful ore deposits, and a Seaway could mainline them to the U.S. mills in the middle of the continent.

Congress heeded the worries of the World War II general-turned-commander-in-chief, and legislation authorizing Seaway construction was signed by Eisenhower in May 1954 with a pen that held a piece of timber from old Fort Detroit, a vestige of the days when the United States and England battled over control of the Great Lakes and the rivers flowing out of them.

Just weeks later, the two countries unleashed an army of 22,000 workers to build seven 30-foot-deep locks on the St. Lawrence River between Lake Ontario and Montreal to replace the hodgepodge of 21 smaller Canadian locks. The United States would build two of the locks. Canada would construct the five others and the costs—ultimately $133.8 million for the United States (to be paid back over 50 years through tolls paid by shippers) and $336.5 million for Canada—were split accordingly. A related project included a $600 million hydropower dam arcing more than a half mile across the river—and directly over the border. The dam, whose cost was evenly split, was also integral to Seaway navigation because it created a 30-mile-long manmade lake behind it that allowed ships to sail over a series of once-impenetrable St. Lawrence River rapids.

Construction crews from both sides of the border tore into the river channel with so much violence and with such heavy earth-moving machinery that they could accomplish in a day what took the earliest, pick-swinging canal builders months, if not years. Just one piece of equipment, known to local school children as the "Gentleman," was a 16-story-high crane with a shovel big enough to scoop more than 56,000 pounds of earth a minute. It was soon teamed with a similarly sized crane called the "Madam." Together they helped make up what was, at that point, the largest concentration of heavy machinery ever assembled on the planet.

≋≋≋

IN THE SUMMER OF 1955, JUST AS THE SEAWAY CONSTRUCTION
was getting underway, a *Newsweek* reporter on the banks of Lake Erie
at the city of Buffalo was left grasping for words to convey the scale of
the project and its prospects to transform such a city. "You can stand
here today and see tomorrow—the multitude of ships flying the flags
of the world, turning the Great Lakes into a Mediterranean and turning
the lake cities into world cities . . ."

The prospects of the Seaway left a *Time* magazine reporter of
the time similarly lathered: "The river and the Great Lakes it drains
will be transformed into a manmade Mediterranean which seagoing
ships can sail westward into North America's heartland. The seaway's
impact on both the geography and economy of the continent will be
enormous. More than 8,000 miles of new coastline will be added to
the United States and Canada. Such lakefront cities as Chicago, Cleveland, Duluth, Buffalo, Toronto and Hamilton will become genuine
deepwater ports . . ."

The reporters were only regurgitating what they were hearing from
Seaway advocates; leaders of every Great Lakes city with a dilapidated
dock were telling constituents that their gritty harbors were about to
be transformed into sparkling international ports rivaling any on the
globe. "The St. Lawrence Seaway will be the greatest single development of this century in its effects on Milwaukee's future growth and
prosperity," Milwaukee port director Harry C. Brockel brayed just
before the Seaway opened in 1959. A downtown Milwaukee store that
spring had already opened a special "foreign shop" to market all the
exotic goods Brockel and other local leaders were convinced would
flood the city docks.

In Detroit, Chrysler was predicting 80 percent of its auto exports
would float out the Seaway, and Minnesotans were convinced the ocean
was about to lap at their state line. "The Seaway will pull Europe closer
to Duluth and away from New York and Philadelphia," editors of the

Winona *Republican-Herald* wrote just after Congress passed the Seaway legislation. "The 'landlocked' Midwest is landlocked no more." And Chicagoans, predictably, saw the Seaway as a chance to shed their second-city status. "The only thing that made New York the biggest city in the country is that everything had to stop there," Robert Kohl, president of the Chicago-based Midwest Steamship Agency, told the *United Press.* "Now there'll be no reason to stop. We'll come right to Chicago with imports and leave from here with Midwestern products for foreign countries."

There was reason for this optimism. Egypt's Suez Canal had changed the way the world works when it opened less than a century earlier, in 1869. The 120-mile manmade waterway connects the Mediterranean with the Red Sea and provided sailors a straight shot between Asia and Europe, trimming about 4,300 miles from the treacherous route around Africa. Today the canal handles about 18,000 ships annually, carrying some 800 million tons of cargo. The Panama Canal further revolutionized global commerce a half century after Suez opened when it cracked the Western Hemisphere in half with a 50 mile cut between the Atlantic and Pacific Oceans that slashed the sailing distance between the U.S. East and West Coasts by about 8,000 miles.

Panama handles around 14,000 vessels annually, carrying more than 300 million tons of cargo, and that volume of cargo is expected to double in the coming years with its recent expansion. Both Panama and Suez continue to be linchpins in global commerce, still hailed as modern wonders of the world.

But the Seaway harbors a much more dubious distinction; it has been said that it stands alone among modern engineering marvels in that it is less famous today than it was in the years before it was built. And the reason: the Seaway locks were built so small they were obsolete almost before the freshly poured concrete could dry.

Even though the Panama Canal was already 50 years old when the first earth was turned on the Seaway, the United States and Canada opted not to build Seaway locks to Panama scale, 1,000 feet long by 110 feet wide. Instead, they decided to build Seaway locks to match those of the smaller, pre-World War I-designed Welland Canal—766 feet long and 80 feet wide. Cost was the reason.

Seaway architects figured that building Seaway locks to Panama scale would be useless unless the Welland locks were expanded as well, and that project alone would cost about $300 million. This would almost double the Seaway price tag, and all but guarantee it would not be funded. In November 1954, just as construction was getting under way, the U.S. Seaway administrator and his staff tried to assure the public that the new Seaway would be plenty big enough. "The majority of general cargo seagoing ships," the Seaway public relations people insisted, "will be able to ply the Seaway when it's completed." And they were right—at that moment.

In May 1956 the Seaway dedicated the new U.S. Eisenhower Lock near Massena, New York, one of the two locks constructed on the U.S. side of the St. Lawrence River. Some 2,000 people attended an event that was covered by both Canadian and U.S. networks. But a bigger deal happened across the state line in New Jersey, just four weeks earlier. That event drew almost no attention, but the world of shipping would never be the same, and the Seaway would never recover from it.

MALCOLM PURCELL MCLEAN, SON OF A NORTH CAROLINA FARMER, had few education or career options when he graduated from high school in the depths of the Great Depression. He took a job pumping gas at a local service station. Three years later he bought a used truck for $120 and went into business for himself hauling dirt for Works Progress Administration road construction projects. Within a few years he was able to buy a fleet of five trucks and pay other men to do the driving. When business sagged a couple of years after that, he was forced back into the driver's seat and began making runs from North Carolina to the New York area. He was stuck in his truck on a Hoboken pier in late November 1937 with a load of cotton bales waiting for his turn to unload when a notion struck. As he watched the stevedores scramble with their cranes and slings, he was not left in awe of their industriousness. He was flabbergasted by all the clumsiness.

Loading a ship at that time was far more an art than the mechanical process it is today. Different products of different shapes, sizes, weights and fragility had to be placed in cargo holds with great care. Some cargoes arriving at the dock had to be held until others arrived so everything could be tucked into a ship's hold just so. It was not unlike packing a grocery bag—the eggs might be the first in line, but they have to wait for the flour and canned soup to be bagged lest they get crushed. This meant that the loading process for a ship in the 1930s sometimes took longer than its voyage across the Atlantic. *There has to be a better way*, McLean thought that day as he sat in his idled truck. It was a thought that would run through his mind over and over, for more than two decades.

On a raw day in late April 1956, McLean was ready to put into action the thought experiment that wormed its way into his brain after that day on the Hoboken dock. Just as Seaway crews were preparing to congratulate themselves for chewing nearly halfway through an enterprise that was, at that time, the largest construction project under way on the globe, McLean quietly, and almost singlehandedly, launched a globe-changing one. He took a run-of-the-mill oil tanker he had named *Ideal X* specifically for his experiment and installed a raised platform on its deck with slots to hold the bodies of 58 trailer trucks that had their wheels removed. "These were not trucks in any conventional sense— the 58 units had been detached from their running gear on the pier and had become *containers*," Brian J. Cudahy wrote in a 2006 report published by the Transportation Research Board of the National Academies. "Arriving in Houston six days later, the 58 trailers were hoisted off *Ideal X*, attached to fresh running gear, and delivered to their intended destinations with no intermediate handling by longshoremen."

McLean calculated the cost to move a ton of cargo aboard *Ideal X* was less than 16 cents, compared to $5.83 per ton for cargo hauled on a traditional ship. The next year McLean converted a World War II cargo ship so containers could be stacked like Legos on top of each other, both below deck and above. The ship was 450 feet long and could carry 266 containers.

McLean's innovation did not change things in an instant. It took years for ship owners, railroads and trucking companies to build the fleets, ports and transfer facilities so the boxes—typically 8 feet wide, 8½ feet tall and either 20 or 40 feet long—could be moved seamlessly between factory, boat, train, truck and warehouse. But what came to be known as the container revolution demanded ever-bigger ships, and by the 1960s the largest container vessels were more than 100 feet wide— already 20 feet too wide for the Seaway's locks. By the 1980s container ships were 1,000 feet long and more than 130 feet wide—50 feet wider than the Seaway locks. Today the biggest container ships are more than twice as wide as the Seaway locks.

ON JUNE 26, 1959, PRESIDENT EISENHOWER AND QUEEN ELIZA- beth, wearing a blue dress with white purse dangling under her left arm, boarded the Royal Yacht *Britannia* to mark the Seaway's completion by steaming through a ceremonial gate at Montreal. The gate was made from the timbers of an old wooden lock that had been built to bypass the previously impenetrable Lachine Rapids, the torrent that had kept so many boats at bay for hundreds of years. The ride up the Seaway that day was pure ceremony—the Seaway had already been open for a couple of months and the leaders of the free world were only on a day trip, and not actually sailing for the Great Lakes. That was a good thing, because what was going on further up the Seaway was a royal mess.

As the first deep-draft overseas ships started sailing for the Great Lakes earlier that spring, the difficulties of operating an elevator system to move 50 million-pound ships 60 stories high were quickly revealed. In the weeks before the first ships sailed in, waters above Niagara Falls remained under a blanket of ice up to three feet thick. By late April the ice in places was still stubborn enough to back up a convoy of 130 ships just below Montreal. Once the ships were allowed to steam inland they just kept bumping into each other.

There were three-day-long jams at the Welland Canal. Jams at Detroit. Jams at Chicago. A blown fuse on a bridge over one of the new canals below Lake Ontario knocked out the bridge's lift and an oil tanker knocked into the bridge, corking the whole Seaway for the better part of a day. A German freighter carrying soybean oil and tallow ran aground in the St. Lawrence River below Lake Ontario. Another tanker hit a St. Lawrence shoal and was almost sunk as it raced for the safety of a dock. A Greek ship got so beat up in Seaway lock chambers that it arrived in Lake Ontario with smashed navigation lights, a busted propeller and a buckled bow. Yet another ship got stuck in the Welland Canal like a semi-truck wedged under a highway bridge. Its captain had to shear off a portion of the ship's bridge to get by. Once clear of the new Seaway locks and the Welland Canal, things in the upper lakes only got tighter because dredging upper lakes' river channels and ports to Seaway specifications would not be finished until the 1960s. A frustrated German captain of 1 of 10 ships anchored near Detroit waiting for the congested docks to clear finally barked: "Didn't you people expect ships?"

The captain of the first American ocean vessel that arrived at

Detroit got so frustrated waiting for dock space he steamed on to Cleveland to pick up other cargo. When he got back to Detroit there was still no space for him to dock. He left without his planned cargo, which included 132 cars and trucks bound for Venezuela. A train ended up taking some of those autos to the coast where they were loaded onto a boat at a port that could handle the job.

"The St. Lawrence Seaway, dream of Midwestern and Canadian shippers for a half century, has now been operating for a month. Experience has been more than a dream, however," one Pennsylvania newspaper editor steamed in May 1959. "In some respects it borders on nightmare."

Things got a little smoother later in the summer as the local pilots got familiar with the channels (Seaway regulations required foreign captains to turn their wheel over to a local sailor on a ship's trip through the system) and the Seaway lock operators got more practice at helping the ships squeeze through the harrowingly narrow chokepoints, but not a lot better. Shippers quickly lost their patience. The Grace Line, one of two U.S. shipping firms regularly serving the Port of Milwaukee during the Seaway's first season, announced at the start of the second season that the Seaway wasn't worth the hassle. The company claimed more than $1.2 million in Seaway-related losses the previous year, in part because of damage ships suffered banging through the locks and channels. Company officials also grumbled the voyage took more than two weeks longer than planned due to bottlenecks at the locks and inadequate port operations. The bad rap spread. The next year, the U.S. deputy administrator of the Seaway blasted American shippers for their "disgustingly small number" of vessels using the Seaway. By the early 1970s, barely 10 years after the Seaway's opening, even some of its biggest supporters were shaking their heads.

"The Seaway—I like to forget it," Dick Miller, information director for the U.S. Seaway for most of the 1960s, lamented to the *Canadian*

Press in 1970, long after it had become apparent that the hoped-for volume of exotic cargoes from foreign ports just wasn't going to enter a navigation corridor best suited for relatively small ships hauling iron in from the Canadian coast and Midwest grain out. "The thing was built on romantic issues—the fourth seacoast and so on," he said. "But you can't romanticize iron ore and wheat."

By 1982 Seaway revenues were so sluggish that Congress forgave the U.S. Seaway agency its $110 million debt, and this was after lawmakers in 1970 allowed the Seaway to stop paying interest on the debt. Without the break, Seaway operators said they would have been forced to increase their tolls by 70 percent, which could have fatally crippled the struggling operation. Five years earlier, Canada forgave its own Seaway agency a debt of about $800 million.

By 1986 the United States stopped charging Seaway tolls, but the volume of traffic Seaway operators sought still didn't come. By 2002 the Army Corps of Engineers reported that the Seaway could only handle about 2 percent of the cargo-carrying capacity of the world's bulk carrier fleet, and 5 percent of the capacity of the world's container fleet.

"The overriding, overwhelming regret is that we built it too small," the late U.S. congressman from Minnesota, Jim Oberstar, a longtime Seaway booster, once told me. "The railroads didn't want to see larger-sized locks in the St. Lawrence Seaway that would compete with the railroads, and the East Coast ports didn't want to see competition from the Great Lakes, and together they combined to limit the size of the Seaway locks."

SHIPPING *WITHIN* THE GREAT LAKES AND ALONG THE SEAWAY and the North Atlantic coast remains a huge business to this day, moving some 200 million tons per year of raw industrial materials like ore, sand, salt and chemicals. And much of it travels through the Seaway locks. But the overseas component of the Seaway's traffic, which peaked at 23.1 million tons in the late 1970s, has dropped in some recent years to less than 6 million tons. Today overseas cargo typically accounts for about 5 percent or less of the overall Great Lakes and St. Lawrence Seaway shipping industry.

Because of the locks' small size, there has long been a push to repurpose the Seaway not as the Great Lakes' gateway to the world, but as a regional navigation corridor in which Seaway boats compete with railroads by ferrying into the lakes containers from East Coast ports. This might make sense on paper, but not on the water—or, more precisely, the ice. The Seaway must shut down for about three months each winter when its locks and channels freeze over, and a navigation route that is shuttered for a quarter of the year cannot compete with trucks and railroads in a world in which businesses manage inventory based on "just-in-time shipping" that demands fast, predictable and perennial delivery schedules.

The U.S. Merchant Marine Academy's Jon S. Helmick once illustrated in a presentation just how choreographed deliveries had become in the 21st-century transportation chain. He pointed to Toyota Motor Corporation's system of delivering engines from Japan to a Kentucky assembly line. "Upon arrival in a Southern California port, the containers are discharged from the ship and then loaded aboard an eastbound double-stack train for overland transport to Chicago. At the terminus of the rail move, the containers are pulled from the train and placed on a truck chassis for the final leg of the journey," he said. "The astonishing reality is that after 17 days in transit, the engines arrive at the plant in Georgetown within pre-scheduled 15-minute delivery windows, at which time they are stripped from their containers and moved directly to the assembly line for installation."

Seasonal closure was a problem Seaway advocates tried to address even before the Seaway opened. One idea was to build a fleet of nuclear power plants to generate enough hot water to keep the waterway ice-free. An earlier suggestion by a McGill University professor to keep ice jams from plugging the St. Lawrence below Montreal involved chemical combustion units loaded into tin containers, and then placing those containers in the river to keep the water flowing. Those thermal schemes went nowhere. Neither did a plan to put a pipe at the bottom of the shipping channel that could pump to the surface ice-breaking bubbles throughout the winter. "Assuming that compressed air were pumped through flexible perforated 1½ inch diameter polyethylene pipe, weighted down and anchored, the total cost of the installation is not expected to exceed two million dollars," reported the magazine *New Scientist* in 1958. "If this comes out, it might be the most spectacular feature of one of the most spectacular engineering projects devised by man."

The U.S. Army Corps of Engineers took up the bubble concept in the 1970s, when it spent some $21 million exploring various ways to stretch the shipping season into the iced-over months. Beyond bubble makers, the agency looked at using a fleet of Coast Guard ice breakers and "ice booms" to shunt the floating chunks away from the locks and channels to keep the water—and ships—flowing. The engineers determined they could indeed keep the upper lakes open for shipping year-round, but the Seaway locks and St. Lawrence River channels would still need to be closed for two months each winter. The concept was scrapped because the cost was astronomical: $451 million, nearly as much as the Seaway construction itself.

Even as the Army Corps scrambled to make the Seaway more attractive by opening it to winter navigation, early generation container vessels were already sailing into the Atlantic port of Halifax, Nova Scotia, and siphoning off Seaway business by working with the railroads serving Great Lakes port cities like Chicago and Duluth.

"There is not a great deal we can do about it," U.S. Seaway boss

David W. Oberlin testified before the U.S. House Appropriations sub-committee in April 1975. He said the only option was to embrace the container revolution and, despite the seasonal closure, try to convert Seaway freighters into small container vessels. It wasn't good advice.

That same year, the port of Duluth invested $2.5 million to install a special crane and related facilities to handle containerized cargo. Duluth did attract three small container vessels in the crane's first year of operation, the port's former director told me. The next year none came. After essentially idling for 18 years, the crane was finally sold to a firm in Beaumont, Texas. The boondoggle that Duluthians had dubbed "the world's most expensive seagull roost" was dismantled and shipped out of town. In a final indignity, the former port director recalled, the crane buyers didn't ship their purchase out via the Seaway.

They dismantled it and hauled it out of town on trucks.

ALTHOUGH GREAT LAKES PORT BOOSTERS CONTINUE TO RUE what might have been had the Seaway been built larger, and had it been engineered to operate year-round, those downstream on the St. Lawrence River still mourn what was lost by its construction.

If you drive down Ontario's King's Highway 2 today just west of Cornwall, across the St. Lawrence River from Massena, New York, you will come across a peculiar road sign. It's a cairn built out of cobble-stones. Atop it sits a brown plank with a yellow arrow pointing south toward the river. It reads: MOULINETTE 1/3 OF A MILE. This is curious. Speed limit signs in Canada were converted to kilometers over Labor Day Weekend in 1977. It's even more curious, because if you follow the arrow, the road leads not to the outskirts of a little town. It dead ends about 100 yards to the south at the banks of a St. Lawrence River bloated by a Seaway dam just downstream. Somewhere under that shimmer-ing blue water rests the remains of an entire town, one once big enough to have its own railroad station, church steeples and service station. No one is sure of the origins of the name; it could be traced to *moulinet*, which is French for "winch," which is exactly what was used to move boats upstream in this stretch of the river when it was a rushing torrent and not a placid manmade lake.

Moulinette and several other towns—including Milles Roches, Dickinson's Landing, Wales, Farran's Point and Aultsville—were flooded in the 1950s to make way for the Seaway. The 6,500 residents in the "inundation zone" were given a choice to be bought out by the government or have their homes moved out of their towns and up to higher ground. Neither option sat well with George Hickey, an 83-year-old retired schoolteacher whom I encountered while taking a driving tour along the Seaway. We met near the banks of the St. Lawrence River where, not far away, 50 years earlier a man from the Ontario power company knocked on Hickey's door and told him and his wife they had one year to move.

But then the manmade flood hit Hickey as fast as those prehistoric torrents that filled the Mediterranean and Black Sea basins.

"The next day I got home from teaching and my wife said, 'We'll be moving tomorrow,'" Hickey recalled with a soft, sad chuckle. They were told that they had been bumped to the top of the moving list and to pack only clothes, nothing more, because they would be housed in temporary quarters during the two weeks it would take for crews to lift the home off its foundation and roll it a few miles up the road.

The new lake swallowed about 38,000 acres, including cemeteries. Sometimes gravestones were pulled up and replanted. Sometimes the dusty bones beneath were left buried under piles of rubble to keep them from washing downstream.

But that manmade flood to make a manmade Mediterranean paled to an entirely different type of flood yet to come, one nobody pondered when all the giant machinery started to chew its way inland from the sea. Almost nobody.

In spring 1955, eighth-grade student Pat Kenney worried about what no one at that time seemed to be worried about. He fretted an ecological disaster might be triggered by reengineering the river in a manner that would connect the once-isolated lakes to the ocean like never before. "I think there should be something done about it for the sake of our freshwater fish," the boy from Bronson, Iowa, wrote Eisenhower as Seaway construction was ramping up. The president passed the boy's concern on to U.S. Seaway boss Lewis G. Castle who, as much as one can in a letter, patted the boy on the head.

"Perhaps you are not familiar with the fact that Lake Superior, which is at the headwaters of our Great Lakes basin, is 600 feet above the Atlantic Ocean. Consequently, the water runs eastward from the Great Lakes area into the mouth of the St. Lawrence River," Castle wrote the boy, "and there is no prospect of the salt-water contaminating the fresh waters of the Great Lakes in any manner whatsoever."

Castle was correct that the river would continue to flow out to sea, but he failed to mention to the boy that the overseas ships sailing inland would carry with them their own mini-oceans. A single Seaway ship can hold up to six million gallons of vessel-steadying ballast water that gets discharged at a port in exchange for cargo. And that water, scientists would learn after it was too late, can be teeming with millions, if not billions, of living organisms.

North America officially got its Fourth Seacoast the next year, and the Seaway boosters were right: foreign cargo *did* flood through the Seaway locks and into the lakes—but it wasn't the type anyone had hoped for. It turns out the Seaway's most important import could not be bought or sold. And it can't be killed.

Chapter 2

THREE FISH

THE STORY OF LAKE TROUT, SEA LAMPREYS AND ALEWIVES

I f the basins that cradle today's Great Lakes were carved by glaciers, and if those massive depressions in the landscape only became lakes after the mountains of ice melted, and if those lakes then evolved isolated from the rest of the aquatic world, then here is an obvious question: Where did all the fish come from?

The answer is that the Great Lakes were not always disconnected from surrounding rivers and lakes. The glaciers that periodically smothered so much of North America during the last ice age waxed and waned over periods that stretched tens of thousands of years. Perhaps it is easiest to think of these pulsing ice sheets as a massive set of frozen waves crashing almost rhythmically, but at geologically slow speeds. Each wave started as the earth cooled to the point that a true summer never arrived in what is today central Canada and the northern United States. Piles of ice would grow snowflake by snowflake, millennia after millennia until they were ocean-sized expanses of ice, the largest of which stretched nearly two miles into the sky and spanned some five million square miles.

Each time one of these frozen waves came creeping down from the north there were freshwater rivers dribbling off its "snout"—the transition zone where the glacier bumped into too much sunlight and too-warm breezes to stay solid year-round. The rivers flowing from these mountains of ice—and the lakes they fed and the grasslands and forests through which they flowed—gave refuge to the plants, animals and fish whose native ranges had been smothered under ice.

Then the climate would turn warmer and the ice sheet would pull back north, drop by drop, mile by mile, and the surviving plants, animals and fish that had been biding their time beyond the frozen zone followed the retreating ice northward to colonize the freshly exposed landscapes and lakes. And then the globe would grow cold again, a new ice sheet would press down into the middle of North America, and the fish and animals in its path would scramble once again for refuge.

Nobody knows precisely how many of these waves of ice plunged down from the north during the last ice age, which began about 2 million years ago and ended only when the last wave retreated barely 10,000 years ago; each time one advanced it scrubbed away the lakes, rivers and altered landscapes left behind by the previous one. But when the last wave pulled north, the massive basins it left behind filled with glacial melt. And these new Great Lakes were eventually populated with the hardy fish and other aquatic species that had been dwelling in the melted waters just beyond the glacier's reach.

The lakes in their early years were constantly shifting in shape, in size and in relation to adjacent waters during the ice sheet's stuttering retreat; the ice sheet began to shrink about 20,000 years ago but would then grow a bit again, and then shrink again in a two-steps north, one-step south fashion. There was a period when the lakes— or their predecessors—were linked to a freshwater sea northwest of today's Lake Superior that was bigger than all of today's Great Lakes combined. There was a time when a river flowed out of what is today Lakes Michigan and Huron and into a massive estuary of the Atlantic Ocean. Another river flowed from the southern end of Lake Michigan into the Mississippi River basin. Eventually these and other early links to waters beyond the Great Lakes basin dried up, and beginning about 2,500 years ago the lakes' sole continuous connection to the outside aquatic world was the river system that roared over Niagara Falls, coursed through Lake Ontario and then rushed down the St. Lawrence Valley and out to the North Atlantic. Species might be able to tumble over the thundering falls and out to the ocean, but the wide-open door for fish and other water-bound aquatic organisms to migrate upstream from the ocean (as well as from other inland waters on the continent) and into the lakes had been shut.

This left the four lakes above Niagara Falls largely separated from the rest of the aquatic world. The lakes might have sprawled across an area half the size of California, but in a sense they were as isolated as a one-acre pond in the middle of a forest until the early 19th century, when construction of the Welland and Erie Canal bypassed the falls and linked the lakes to the Atlantic Ocean. Pulling the Niagara plug that had protected the lakes for millennia triggered an ecological calamity best illustrated by the rise and fall of three species of fish— lake trout, sea lampreys and alewives. Their story shows how a delicate ecological tapestry that had been thousands of years in the making unraveled in just a couple of decades.

AT THE BASE OF THE FOOD WEB STITCHED TOGETHER AS THE upper lakes' river pathways to the outside world closed to migrating fish were phytoplankton. That plantlike life was gobbled up by tiny floating animals (zooplankton), which, in turn, provided meals for things like mollusks and crustaceans. Upon those larger critters feasted bait-sized fish like the ninespine stickleback, slimy and deepwater sculpins, as well as a minnow called the emerald shiner. The little fish sustained the lakes' medium-sized fish, including perch, their larger cousin the walleye, and smallmouth bass. Also in the lakes' shallower areas lurked the giant sturgeon, which can live more than 100 years and grow to seven feet by rooting on the bottom for aquatic insects, crustaceans, mussels and the occasional small fish.

To call all this a food chain is an oversimplification. What gobbled what was not strictly linear; baby predators were often a meal for the same prey fish that those predators would have eaten had they survived to adulthood. Some fish species feasted on their own kind, and some of the big fish in this Great Lakes' web of life didn't favor little fish but instead competed with them for crustaceans and plankton. Whitefish, for example, could grow more than two feet long and swell to over 10 pounds on a diet that consisted mostly of bottom-dwelling organisms and crustaceans that migrated nightly up from the lakebed under the cover of darkness in search of plankton. In the same family as white-fish and occupying similar strands in the food web were more than a half-dozen closely related schooling species. The most well-known of these was commonly called lake herring but there were also sev-eral types of "chubs" with peculiar monikers like bloater, shortnose, blackfin, shortjaw and kiyi. Some of these foraging fish, collectively known as ciscoes, lived like mini-whitefish, schooling up to chase the crustaceans that migrated nightly off the lake bottoms. And chasing all these clusters of little fish were the yawning jaws of the lake trout—the spider in the Great Lakes' web of life.

The giant trout that can grow to a wolf-sized 70 pounds are in the same family as salmon but although the fish share the same ancestry, in many ways they could not be more different. Salmon hatch and spend their youth in freshwater rivers and streams and then descend to the ocean to feast for two or three years before returning to their native freshwaters to spawn once and die. Salmon can reach their maximum size—sometimes approaching 100 pounds—in just a few years by devouring schooling prey fish like herring. But salmon are a picky breed. If they can't find enough little schooling fish to eat during their short life, their own metabolism will burn them up; they will starve or become weakened and die of disease.

The lake trout, one of the hardiest and most recognizable native species in the Great Lakes.

Slow-growing lake trout are a beast of a different sort. They can live for decades, reproduce year after year, and are able to grow fat in the same conditions that would starve a salmon. Having evolved over millions of years to survive the frigid, relatively sterile glacier-fed rivers, lake trout eat just about anything they can find in the lakes—from plankton to insects to other fish—and are also nature-built to weather long periods of famine typical of such waters.

If their food supply hits a low cycle, lake trout simply throttle down their metabolism and stop growing as they wait out the lean years. Were a trout to let its guard down in this fashion in the ocean, a bigger fish likely would swallow it whole. But adult lake trout in the Great Lakes food web only had to worry mostly about eating, not being eaten. This ability to pace its growth with available food sources made it the perfect fish to regulate—or, more accurately, harvest—the slow flow of energy through the Great Lakes that starts 93 million miles away.

"There is nothing like them," says Mark Holey, a Great Lakes trout specialist with the U.S. Fish and Wildlife Service. "Lake trout have the characteristics to most efficiently transfer the energy in that ecosystem from, basically, sunlight into fish flesh."

And just as humans bred canines for distinct characteristics—shepherds for herding, hounds for hunting, Dobermans for protection, etc.—a similar sort of specialization happened naturally with lake trout in the Great Lakes. Because of the relatively small number of predators that made their way into the lakes before the rivers connecting them to the outside world disappeared, lake trout evolved in the Great Lakes over thousands of years to fill multiple ecological niches that otherwise might have been occupied by other species. In Lake Michigan alone some biologists believe the fish were organized into at least 100 stocks, many of which became isolated populations that bred only among themselves. Across the Great Lakes, each stock became uniquely adapted to thrive in the areas it colonized. Some populations thrived amid mudflats or boulder-strewn open waters hundreds of feet deep by accruing loads of fat in their muscles and body cavities that gave them a buoyancy so they could swim easily throughout the varying lake depths, whether near the surface or in the high-pressure zones of the deep. Their eyes were larger and positioned closer to the top of their heads than some other forms of lake trout—perfect for a fish that, shark-like, attacked schooling whitefish and ciscoes from below.

Other breeds lived in shallow waters and competed with whitefish and ciscoes for crustaceans and insects. Some were better equipped to strip plankton from the water. Some spawned in cobble, others on rocky reefs and still others in areas where the lakebed was smothered in algae. There were also some stocks that spawned, salmon-like, in rivers and streams.

Depending on where it lived, a trout's mature size varied from barely 12 inches to nearly four feet. Its skin color could be green or brown with patches of yellow or orange. Or its skin could range from white to nearly black, and span all the grays and silvers in between. Depending on the stock, its flesh ranged from white to pink to deep red. Most lake trout stocks spawned in the fall but others spawned in summer and still others in spring.

The early fishermen who chased these fish gave the stocks different names. There were red fins and yellow fins. There were buckskins, grease balls and paper bellies. There were moss trout, shoal trout and fats. There were bay trout and there were black trout. And they were legion.

The recorded natural history of all these lake trout stocks is sketchy; overfishing beginning in the mid-1800s began to take its toll before a comprehensive stock survey was conducted. But an early attempt to explain the various populations in northern Lake Michigan was made by one James J. Strang, a fiery rival of Brigham Young for leadership of the Mormon Church after the religion's founder, Joseph Smith, was assassinated by a mob in Carthage, Illinois, in 1844. Brigham Young took his faithful to a relatively verdant valley in the Utah desert, which today is known as Salt Lake City—home to a worldwide church claiming more than 15 million members.

Strang, more latter-day pirate than saint, took his own flock in the opposite direction—eastward into the middle of northern Lake Michigan. There, on the 13-mile-long, 6-mile-wide Beaver Island, the 5′4″ Strang, with a beard as orange and long as a carrot, fashioned himself a crown, stuffed some cushions with island moss and called it a throne and proclaimed himself king.

"For any little disobedience of his harsh laws, he ordered floggings," recalled Stephen Smith in a 1940 newspaper story. Smith, 91 at the time, claimed to be the last person alive to have lived under Strang's brutal rule. "He killed any number of men and women, and had others tied up and flogged 'til they bled. He sent men to loot gentile (non-Mormon) stores and even had pirates sailing around the island to rob the fishing boats."

Like his rival Brigham Young, Strang took multiple wives, including one who dressed as a man in a black coat and stovepipe hat, called herself Charles Douglas and claimed to be Strang's "personal assistant." During his six-year reign Strang survived a naval battle with mainlanders as well as a trip to U.S. District Court in Detroit, where he was accused of counterfeiting, piracy, and interfering with the mail and murder, among other charges.

"He talked to that jury and his tongue was like silver. And that jury believed him and said, 'Not Guilty' to all charges against him," Smith recalled. "King James came back to Beaver Island more full of himself than ever, even the U.S. Government couldn't beat him."

But the man Smith called a "cocky little tyrant" was not all trouble. He had so many followers in his church—up to 12,000 at its peak—that he was able to get elected to the Michigan Assembly in Lansing, where by all accounts he acquitted himself well as a lawmaker. He established a newspaper. He was an abolitionist who granted blacks full membership to his sect more than a century before mainstream Mormons did.

And he became a self-styled naturalist who was among the earliest to attempt to classify the types of lake trout swimming in the waters off his island. In an 1853 report he sketched the life history of a plump trout known as a siscowet, which, because of its white flesh, he said some fishermen (incorrectly, it turned out) speculated may be a "mule"—a cross between lake trout and whitefish. He also made note of the skinnier but larger "Mackinacs" that lived in shallower waters, swam alone except when spawning, and gobbled up everything under the surf, regularly plundering the nets fishermen had set to catch schooling whitefish.

"They are a voracious fish of prey, seizing and devouring so far as we can learn, every other kind, even their own," Strang wrote. "Herring are their constant prey. Whitefish of two pounds weight have been found within the belly of the trout. Small trout are sometimes found in them." Strang explained that catching lake trout at that time, which he noted could grow to more than 50 pounds, was a ridiculously laborious process, especially in winter. "The moment the bite is felt the fisherman throws the line over his shoulder, and runs with all his might, in a direct line, till the fish is on the ice," he reported.

The trout weren't much easier to catch from a boat. Strang described how fishermen let the fish pull the boat, *Jaws*-style, until it exhausted itself. This was no easy way for a fisherman to make a living. He reported two fishermen working together full-time did well if they caught 800 pounds of trout in a week.

Strang was shot in the head in 1856, according to Smith, by two disgruntled followers who had left his church after they refused Strang's edict that their wives—along with all other women on the island—wear bloomers. The murderers were never charged, and Strang's tyrannical reign was largely lost to history.

Soon so too would be the era when a couple of fishermen's weekly 1,000 pound lake trout haul would be considered huge.

In the years following Strang's death, Great Lakes fishing evolved from a local industry that sustained lakeshore communities into a treasured national resource as the fishing fleets became motorized and the net hauling mechanized. The annual lake trout haul on Lake Michigan alone by the 1890s was topping more than 8 million pounds. Fishermen on Lake Superior and Lake Huron reported somewhat smaller but similar hauls, and harvests on Lake Ontario exceeded one million pounds. (Lake Erie, a much shallower and warmer body of water, had a much less significant lake trout fishery, though a stunning 44 million pounds of ciscoes were hoisted annually by 1890.)

No matter how hard the lake trout stocks were fished, the lakes continued to yield millions of pounds of lake trout annually, decade after decade all the way into the 1940s, when some 100 million pounds of all species of Great Lakes fish were being harvested each year. And then, in just a matter of several years, stocks of lake trout and several species of ciscoes suddenly vanished. Whitefish populations across the lakes were similarly decimated, if not completely destroyed.

Greedy fishermen alone could not have wrought such instant devastation. It turned out they had an accomplice in an environmental calamity that was unlike any in the history of freshwater fisheries—a stealthy, eel-like bloodsucker that wriggled up the shipping canals built in the 19th century that had destroyed the lakes' natural barrier to the East Coast. This ancient predator exposed the young lakes for what they were—ecological babies, really. And just as vulnerable.

The speed and extent of the fishery collapse that followed the sea lamprey's discovery in Lake Michigan in 1936, in Lake Huron in 1937 and in Lake Superior in 1938 left ecologists and fishermen baffled. By 1949 federal biologists were predicting the "complete" collapse of lake trout stocks on the three lakes, and whitefish and ciscoes were headed in the same direction. This is how one newspaper reporter in 1950 described an ecological meltdown of unprecedented scope:

A few weeks ago Henry Smith, commercial fisherman of Waukegan, Ill, took his boat out into the trout beds of Lake Michigan. He set four miles of nets.

Several days later he went out again and lifted them. He caught six trout. Five years ago the same operation might well have produced 6,000 pounds of fish.

Now those great succulent trout are gone. The Great Lakes' fishing communities are crumbling. Millions of dollars' worth of nets and gear and boats lie useless. Young men seek other jobs, but the older ones hold on, desperately trying to eke out a living catching coarser fish.

There is a murderer abroad in our Great Lakes that has all but destroyed one of America's greatest commercial and sporting fish.

His name is the sea lamprey, an eel-like bloodsucker originally a native of the Atlantic ocean. Having begun on the lake trout because of its small, soft scales, this same killer is now preying on the whitefish, the herring, the chub . . . anything that moves.

Indeed. Speaking later that same year at a gathering of the American Association for the Advancement of Science, a Michigan professor revealed that the lampreys had begun attacking humans. "However," explained Wayne State University's Charles Creaser, "they do not try to feed as they do on fish. They release their hold when the swimmer leaves the water, leaving only a tooth pattern on the unbroken skin."

Creaser also noted the lampreys had begun attaching themselves to motorboats traveling as fast as 15 miles per hour, a phenomenon he speculated likely contributed to their fire-like spread across the Great Lakes throughout the late 1940s. In Lake Michigan alone the annual lake trout commercial harvest in 1944 was still nearly 6.5 million pounds. Five years later it had dropped to 342,000 pounds, and five years after that, it was zero. It was a similar story for whitefish, of which Lake Michigan commercial fishermen harvested nearly 6 million pounds in 1947. A decade later the annual catch had dropped to 25,000 pounds. Trout crashes also happened on Lakes Huron and, later, Superior and, to a lesser extent, Lake Erie, which not only had a smaller lake trout population but also lacks the cold, fast-flowing crystal clear spawning rivers required by what is, pound for pound, one of nature's most devastating—and durable—predators.

THE MOST GHASTLY THING ABOUT THE FOSSILIZED BEAST THAT the young paleontologist chiseled from a nearly 400-million-year-old rock in 2005 was its mouth. The ancient fish, discovered in a pile of shale near the southern tip of Africa, had a sleek body and a fat head, like a giant oversized tadpole. Atop that head yawned a single nostril. The little creature's two beady eyes were set back from the front of its face and pushed to the side in a manner that made it apparent that this was a killer that didn't take its prey head-on. And then the mouth. It really wasn't a mouth at all, in terms of what you think of when you think of lips and jaws and teeth. It was just a round hole in the bottom of its head rimmed with 14 fangs in a manner that created the most absurd, exaggerated overbite imaginable. If Bart Simpson had a pet water snake, it would look something like this. It was hard to imagine how this jawless creature could chew and devour its prey, and it couldn't.

The find of this ancient sea creature, which opened a window into life in the prehistoric oceans was, oddly, a byproduct of the civil unrest that plagued South Africa in the waning days of apartheid. A main road slicing through the slums of Grahamstown, a city with a population of about 70,000 in East Cape, was being rerouted in the mid-1980s in order to keep whites and blacks from having to cross paths. The construction project involved cutting through hillsides heavy with vegetation on the city outskirts, exposing a wall of rock cut through by black shale. Rob Gess, then a teenager in the area with a keen interest in geology, started picking through those flaky shale scraps with a pocket knife after he found he could separate the layers of shale as if he were turning the pages of a book. And, thick with the fossilized remains of plants and fish, those pages told the story of what life was like in one ocean lagoon 360 million years ago. The shale, it turned out, was once muck in a swampy lagoon that existed at a time when Africa, South America, Antarctica, India and Australia were fused into the super continent called Gondwana.

In the late 1990s when the freshly exposed shale faces along the new roadside started to slough, the government provided Gess, by then a graduate student in paleontology, a flatbed truck and six laborers to remove about 30 tons of the material. Gess eventually built a shed to keep all his free, fossil-rich rock from weathering. It was there in 2005 that he discovered the remains of the predator that evidently thrived by latching onto the bellies of its victims with its suction-cup mouth rimmed with teeth made from a protein similar to that found in human fingernails. A toxin in its saliva acted as an anti-coagulant to keep the host's blood flowing, and the creature hung on until its own belly was full—or until there was nothing left for its victim to give. It was a most primitive way to make a living—and a killing—but one that proved devastatingly effective.

Gess's specimen showed no evidence of a bony skeleton, only a head, gill basket and spine made of cartilage—squishy material that generally makes for a poor donor to the fossil record. This meant the fossil was actually only an impression unearthed between the flakes of rock, almost as flat as a picture. So how could the scientists determine from this sketchy image so much about how this creature lived and how it attacked its prey? It turned out this crude killer had left behind other hints of how it lived—its babies. And those babies' descendants left behind more offspring. And so on, through the Carboniferous, Permian, Triassic, Jurassic and Cretaceous periods and into the Ceno-zoic period that began nearly 66 million years ago, and through all its epochs—the Paleocene, Eocene, Oligocene, Miocene, Pliocene, Pleisto-cene, Holocene, and on and on and on. All the way up until the day the Erie Canal opened—in 1825.

Gess's fossil represents a direct ancestor of today's sea lamprey, which is native to the Atlantic Ocean's coastal waters and the rivers that feed them. Aside from some minor evolutionary tune-ups, includ-ing a few more rows of teeth and a little bit more length and girth, Gess's find showed the predator has not evolved significantly since the day its ancestor got stuck in the mud of that Gondwanan lagoon. Con-ventional wisdom is that animals with such a specialized design are likely to fade as their prey evolve or go extinct. But not lampreys, which somehow managed to survive four of the earth's five mass extinctions. They outlasted placoderms, which could grow larger than the biggest sharks of today and were protected from razor-toothed predators with turtle-like shells on their backs and heads, as well as the shark-sized reptiles called ichthyosaurs that swam the ocean a quarter billion years ago. And they survived plesiosaurs, marine predators that could grow to a dinosaur-sized 50 feet in length.

"Essentially," Gess said, "this is such a successful morphology that as long as there are cool waters and aquatic vertebrates for it to feed on, the lamprey continues to be a success." And nowhere has that success been greater than in the Great Lakes. Sea lampreys today are but a bit player in the Atlantic Ocean's overall ecology. They have endured, but haven't decimated ocean stocks. They also give back to the food web; lampreys are a welcome meal for fish species like cod, swordfish and striped bass.

But the saltwater native proved to be an ecological menace when it finally made its way up the canal system from the ocean and into the Great Lakes, whose vastness above Niagara Falls belied the relatively simple food web that existed before construction of the Welland and Erie Canals. Four species of lampreys are native to the Great Lakes basin and have existed for thousands of years in harmony with the other Great Lakes fish. Two of the species never reach the fish-attack stage and instead spend their lives burrowed into streams that feed the lakes, like worms. They survive by poking their heads from the streambed and sucking plankton and other microscopic material from the water flowing past. The two other native Great Lakes lampreys do prey upon fish in open waters, but are only about a foot long and never posed a threat to the existence of other Great Lakes fish species.

And then, with the expansion of the Welland Canal toward the end of the 19th century, came the sea lamprey. Like salmon, sea lampreys are anadromous, which means they spend the first part of their lives in freshwater rivers and streams before descending to the ocean to live as predatory adults and then return to freshwater streams and rivers to spawn and die. But, also like salmon, sea lampreys don't necessarily need a saltwater component in their lifecycle—not if they can find a body of water big enough and filled with enough fish to serve as a surrogate sea.

THE FIRST GREAT LAKES SEA LAMPREY WAS DISCOVERED IN Lake Ontario in 1835. There is debate as to whether sea lampreys had always existed in Lake Ontario, but many researchers believe the lampreys only colonized the lake after the Erie Canal opened. The theory is that lampreys, which are native to East Coast rivers, including the Hudson River, could have ventured into the eastern portion of the Erie Canal and then invaded Lake Ontario by swimming up the Erie Canal's "feeder" canals. These manmade waterways descended from the Lake Ontario basin to keep water flowing in the eastern portion of the Erie Canal. Because adult lampreys are built to swim upstream, the water coursing down from Lake Ontario would have drawn them in that direction.

The sea lamprey, which devastated the lake trout population soon after it invaded the Great Lakes.

It might seem logical that at least some lampreys would have continued their westward migration toward Lake Erie as well, but today's lamprey experts find that unlikely. The water in the western portion of the Erie Canal was too warm, too slow moving and too polluted, especially compared to the alluring cold, clean waters pulsing into the Erie Canal from the Ontario basin. But this does not explain why lampreys did not immediately make the jump from Lake Ontario to Lake Erie by swimming up the Welland Canal when it opened in 1829—nearly a century before the first lamprey was found above Niagara Falls. One explanation is that Lake Ontario didn't have the appropriate spawning habitat for sea lampreys to thrive to the point that they began to crowd each other out and had to seek new waters to invade. Another possible explanation is the design of the early Welland Canals; lampreys seeking upstream waters to spawn would have been flummoxed once they hit the middle of the canal where the water did a most unnatural thing; it flowed in both directions—toward Lake Ontario and toward Lake Erie. Because of a rocky high point on the canal route that put the canal bed at an elevation higher than Lake Erie, water from a feeder canal had to be channeled into the Welland Canal near this crest so boats could float over it. When water from that feeder canal coursed into the Welland Canal at the crest, it then flowed in two directions in the Welland— south toward Erie and north toward Ontario. So a lamprey, nature-built to always swim upstream, would have sensed the downstream switch in current at the crest and headed into the feeder canal and, the theory goes, likely would have ended up in a tangle of upland streams and soggy ditches draining agriculture lands—a biological dead end and, if the story ended there, a life saver for fish in the upper Great Lakes.

This underwater hump in the Welland Canal—perhaps the last line of defense preventing a sea lamprey invasion of the upper Great Lakes—was obliterated when the canal's third expansion was completed in the 1880s and with upgrades in the following decades. The deepened channel finally allowed Lake Erie water to flow continuously down the Welland Canal, through the locks, and into Lake Ontario. This gave the lampreys a continuous upstream migration into Lake Erie.

It is also possible that lampreys had been trickling into Lake Erie ever since the first Welland Canal opened—and perhaps hitchhiking their way through its waters by latching onto boat hulls—and that it just took years, decades even, for a breeding population to become established in Lake Erie and then grow large enough for humans to take notice.

Whatever the reason, the first sea lamprey was not found above Niagara Falls until 1921, a 21-inch adult taken in open lake waters about 200 miles beyond Buffalo on the western end of Lake Erie. It would be another 15 years until a lamprey was discovered in the next Great Lake, when Lake Michigan commercial fisherman Frank C. Paczocha found a 15-inch specimen latched just beneath the eye of a four-pound lake trout caught in the waters off Milwaukee. I can only imagine what went through his mind as he stared for the first time at a creature that looks as if it is not of this world and, given its primordial lineage, it isn't.

In spring of 2015 my daughter's sixth grade science class carved into a batch of lamprey carcasses, probably because the dissection specimen company was selling them cheaper than standard-issue frogs. It's one thing to see a picture of the prehistoric parasite, and quite another to run a finger over its dark gray, scaleless skin and, lightly, across its pin-sharp teeth. The foot-long specimens that arrived vacuum packed had a pencil-lead sized nostril on the top of their heads and black, bulbous eyes, behind each of which trailed seven eye-like slits—the lampreys' gills. Unlike most fish that take water—and the dissolved oxygen it contains—in through their mouths, lampreys pull their oxygen through these oval openings because their suction-cup mouths are built to be latched onto prey around the clock.

I was invited by the middle school science teacher to attend one of these squeal sessions, which peak when the 12-year-olds use scissors to cut through their specimens' skin to expose a pale orange paste that is tens of thousands of orange poppy seed-sized eggs. "These are its babies?" howled one girl. "Oh, that is nasty."

Those eggs, had they been allowed to hatch in the wild, would have grown into adults up to two feet in length and peaked in weight at about a pound. It is an ecologically pricey pound—each lamprey can kill 40 pounds of fish during the year or so it spends chasing its prey.

The problem in the early days of the invasion was that the young Great Lakes had no natural predator equipped to control a killer so stealthy and so fiendishly efficient that the press had taken to calling it a vampire. But then a University of Michigan graduate student and World War II veteran named Vernon Applegate showed up and did what no creature in the past 360 million years had apparently been able to do. He got under the lamprey's skin. He figured out how it migrates and how it hides. How it feeds, how it breeds, and how it dies.

And then he put a stake in it.

BY 1950 THE GREAT LAKES SEA LAMPREY INFESTATION WAS AT ITS peak, its commercial fishing industry at its low point, and the lakes' intricately stitched together ecosystem, thousands of years in the making, was in shambles. It was also the year Applegate, after three years of field research in some of the most lamprey-infested waters of the Great Lakes, produced his dissertation, *The Natural History of the Sea Lamprey*, Petromyzon marinus, *in Michigan*. The 334-page document was fat with charts, pictures, sketches and graphs, and it evidently left at least one of his University of Michigan professors rapt. The chief of the university's Institute for Fisheries Research made a point of adding a page inside Applegate's dissertation cover noting it was a document for the ages, that it should be sent straight to the Library of Congress for microfilming and to be catalogued. "Unusually and exhaustively detailed" is how the fisheries chief termed the work that included exquisitely detailed information about the lampreys' physiology as well as a litany of statistical breakdowns on everything from the time of day lampreys choose to swim upstream to reproduce to the types of rocks they use to build their spawning beds to whether the female or male does more work in that pre-mating enterprise. (It's the male.) But the reviewer also called the document "clear and pleasing." Pleasing, because Applegate was clearly on his way to figuring out how to get under the species' ancient, slimy and theretofore impenetrable skin.

Applegate, a wiry ex-infantryman from Yonkers, lived for three years along two lamprey-infested rivers in northern Michigan in search of a weakness in the lifecycle of one of evolution's most durable models. He did it with an intensity that, more than a half century later, still leaves those who worked with him—or who had brief encounters with him—bemused. Applegate toiled around the clock, chasing the slithering gray or black parasites up rivers through the night and into dawn with flashlights and a notebook. He set traps to catch adults swimming upstream to spawn, and traps to catch young lamprey riding downstream on springtime floods toward the lakes. He built outdoor pens to watch them breed. He peeped into their evolutionary secrets through the glass of the aquariums at his lab on the shore of Lake Huron, whose ecological health evidently became more important to him than what was going on under his own pale skin.

"I've heard him described as living on cigarettes and aspirin," said Howard Tanner, a renowned Great Lakes fisheries biologist, who met Applegate while Tanner was studying at Michigan State University, and who would one day, if unintentionally, undermine Applegate's goal of restoring the lakes' native lake trout. "He was very intense. A small man. Red haired."

A perception by many early on in the Great Lakes invasion was that the lampreys devastating the fish populations were primarily a big-water inhabitant. But it turned out their killer stage comprised

only a snippet of their life. The majority of a sea lamprey's existence—
more than five years of its roughly seven years—is spent as a blind,
worm-sized vermin burrowed into the beds of rivers and streams
feeding the Great Lakes. Only their mouths are exposed for them to
suck from the water algae and any other nutritious material drifting
on the stream currents. That meant that for every crop of lampreys
swimming in the lakes' open waters—and there were hundreds of
thousands of them by the time Applegate began his studies—there
were maybe six times that number burrowed into streambeds, growing ever so slowly, invisibly and inexorably, until it was their turn to
attack.

"Aquarium observations make it easily understandable why lamprey larvae are seldom observed in their 'beds' in a stream. The vibra-
tions set up by footsteps across the floor of a wooden building caused
all aquarium-held specimens to retreat from the surface into the depths
of their burrows," Applegate wrote in his dissertation. "After several
minutes, if all remained quiet, they returned to the surface again and
resumed feeding."

Applegate noted that footsteps along stream banks triggered a similar response and "for this reason, individuals of this life history state are
seldom seen, even by careful observers, in their natural surroundings."

Adult lampreys returning to the rivers and streams to spawn
were similarly cryptic. Applegate noted that when researchers stepped
into a stream where pre-spawning adults were lurking under rocks,
logs, overhanging banks and in the depths of dark pools, their escape
instinct was extreme. "When prodded from these hiding places they
dash blindly away with little regard for the direction taken," he wrote.
"In several instances, disturbed specimens darted at right angles to the
current with so much force that they slithered several feet up onto a
low, grass bank or mud flat."

Biologist Vernon Applegate holding a lake
trout under attack by a sea lamprey.

Applegate learned that spawning lamprey preferred streams with
bottoms peppered with gravel that had a diameter no smaller than
three-eighths of an inch and no bigger than two inches, and that they
typically did not migrate up those streams until early spring, when the
water temperature rose above 40 degrees. He found in one creek that
he studied for three years the upstream migration took place under the
cover of darkness up to 99 percent of the time. Thousands of lampreys
could infest a single stream like a virus, invisible to everyone—except
Applegate. He watched how individual male and female lampreys work
together to build spawning beds by excavating stones and moving
them about with their suction-cup mouths (hence their Latin name,
loosely translated as "stone sucker"), stacking them just so to give the
fertilized eggs drifting downstream on the current a protected place to

settle. He found the nests by learning precisely what type of material they needed to successfully reproduce. Gravel with a little sand was ideal. Boulders, bedrock and rubble were deal killers. So were streams that were entirely soft-bottomed. Applegate also learned that lamprey needed a specific type of stream current, one strong enough to keep the eggs floating downstream but not one so swift it would sweep the eggs beyond the newly built nests.

Like a private detective trying to find out a subject's sexual habits, he staked out the stream banks on one stretch of river in 1948 and found it had 954 nests. He then observed actual sex on 338 of those spawning beds. He found 71 percent of those beds were home to monogamous couples; 13 percent were nests with one male spawning with two females, and he found nine nests that had one male and five females. He described the actual spawning act with a crisp clinical detail that echoed that of the work being done at the same time by Alfred Kinsey at the Institute for Sex Research across the state line at Indiana University.

"The male approaches the female generally along the long axis of her body which is parallel to the current. In doing so, he frequently runs his mouth lightly over the anterior half of her body until the branchial zone is reached," Applegate wrote. "At this point the male fastens himself firmly to the female with his mouth. Almost immediately he wraps the posterior third of his body in an abrupt half-spiral about that of the female so that their vents are approximated. The extrusion of the eggs and milt is preceded and accompanied by a very rapid vibration of the bodies of both individuals for a two- to five-second period. Following that, the male releases the female immediately." Applegate also kept a keen eye on the thermometer and noted none of this occurred until a stream warmed to 50 degrees.

Now he knew where, how and when his subjects spawned. He had cages built in streams so he could contain a lamprey couple through the whole reproduction session. He watched the freshly fertilized eggs coast downstream until they were caught in the lower wall of the spawning bed. Then he watched the parents anchor themselves on rocks with their mouths, upstream from the spawning bed, and violently shake their bodies to kick up sand that floated downstream to give their young a protective cover in the stone wall.

The parents would then take a breather for a few minutes and repeat the breeding process. Over and over until both were, in the most literal sense, spent—both inevitably died soon thereafter. The mating period typically lasted about 16 hours, though one couple he watched in early 1947 mated for three and a half days. Spawning females release as many as 100,000 eggs, but Applegate noted that usually fewer than 1,000 of those eggs hatch. That normally didn't happen for 10 or 12 days after the spawn. Around the 20th day after fertilization, the squiggly lampreys, smaller than a sliver of finely shredded cheese, emerged from the pebbly home their parents had built for them and drifted down stream until they hit calm waters—eddies, side channels and wide spots in the streambed—where soft sediments are found. There, they plunged to the bottom, where they would make their home for the next five or six years.

If the current failed to steer the juvenile lampreys to welcoming, soft-bottomed streambeds, Applegate's lab experiments revealed just how determined larval lampreys could be. He noted in his aquarium experiments that if a tiny burrowing lamprey hit a rock while diving for a soft place to hide, it often knocked itself out. But when it regained consciousness, it almost invariably slithered off in search of softer material into which it could find a home.

After lurking and feeding in the streams for a half-decade or more, growing on average only about an inch per year, the blind, burrowed lampreys would begin their transformation into bloodsuckers. Eyes emerged on the side of their heads. So did the creatures' horrifying circular mouths with spiky rows of teeth and piston-like tongues, rough as an emery board to rasp away their victims' skin and scales. Once the metamorphosis was complete, the new crop of lampreys erupted almost all at once from the streambed and headed for open waters. Applegate noted most of the migration came in late March and early April, as water temperatures rose, but before the thermometer hit 41 degrees.

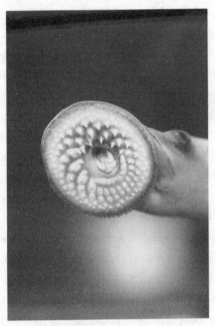

The sea lamprey uses its suction-cup
mouth to latch onto its victims' bellies.

"One of the most striking characteristics of the downstream movement of newly-transformed sea lampreys is the abruptness with which large numbers of individuals suddenly leave the mud banks and move downstream," he wrote. "Under the impetus of rising waters, a virtual emergence takes place and hordes of the new adults travel downstream on the rise and crest of floodwaters. This surge of movement downstream frequently ends as suddenly as it begins."

The lampreys' open water fish-feasting period lasts 12 to 18 months during which they range as far as 200 miles from their spawning streams. Unlike a species like salmon, lampreys don't have a homing sense for their birth waters; they'll swim up any stream their keen noses tell them has a larval lamprey population.

Applegate's research suggested that the slithering killers marauding across tens of thousands of square miles of open Great Lakes water weren't indestructible after all. They were sitting targets, if you knew when and where to shoot. "Plainly," Applegate concluded, "the most vulnerable times in the lamprey's life are its periods in the stream—as a larva or young migrant and later when it goes back to spawn."

The initial control strategy, which was already in use before Applegate presented his dissertation, was to attack the lampreys on the Great Lakes' most heavily infested streams and rivers by building weirs, which are mesh barriers that allow water to flow downstream but block the upstream passage of spawning adults. The concept worked. Three months after Applegate defended his dissertation, the U.S. Fish and Wildlife Service reported that experimental traps in 12 northern Michigan streams had caught 29,425 adult lampreys set to spawn. This was a remarkable early success; if one lamprey could kill 40 pounds of fish, those dead lampreys alone could have been responsible for killing some 1.2 million pounds of native fish. And since each female lamprey could produce several hundred offspring, the biologists were getting confident they were on their way to throttling the explosion.

Researchers had also begun to explore electric barriers, a strategy that would prove effective in deterring upstream migration of adults but ineffective on downstream-migrating juveniles that, even if incapacitated by electricity, could still drift through the barrier zone on swollen spring currents. The electric barriers were also costly to operate and prone to failing. The biggest problem with both weirs and electric barriers was that they could be breached with floodwaters. Also, because a creek could have six year classes of lampreys burrowed into its streambed at any given time, the barriers would have to be successfully operated for the better part of a decade. If one failed just one time, in just one year, it would unleash a fresh crop of parasites to colonize the lakes anew.

Applegate knew that something more effective than these physical barriers was needed. And he knew he was in a race against time—against a creature for which time didn't seem to pass. He was determined to control the lamprey infestation in time to save the remnant stocks of Lake Superior lake trout that had up to that point managed to survive the infestation. He hoped eventually to use them in a breeding program to restore lake trout populations to Lakes Michigan and Huron. But he worried that it might well be too late. Even if government crews were able to install weirs on all the spawning tributaries to Lake Superior immediately, there would still be several years in which larval lampreys would continue descending into the lakes, and that might be all it would take to doom the Lake Superior population of lake trout as well.

He decided that he needed a poison that would destroy the lampreys without wiping out all the other species in the streams or the lakes they fed. His goal was ambitious—"Complete eradication of sea lampreys above Niagara Falls."

The problem was, such a poison did not yet exist.

IN A CONVERTED COAST GUARD LIFESAVING STATION AT THE northern tip of Michigan's lower peninsula in the early 1950s, the freshly graduated Applegate, who had taken a job with the U.S. Fish and Wildlife Service, began a secretive program to develop the perfect poison for juvenile lampreys. One of his colleagues was Louis King, who still remembers the despair he felt the day he arrived on the shore of Lake Huron, fresh out of graduate school in Missouri and with a wife and children in tow. He said the lake looked dead.

"What I saw when I got here," the 84-year-old said in the living room of his home in far northern Michigan near the shore of Lake Huron, "was virtually a desert. A big desert. Nothing was there. No commercial fishing. No recreational fishing." King said he was "overwhelmed" by the scale of the waters, which seemed to him to be more ocean than lake. "I thought: How in the world could they ever, ever control lamprey in this vast body of water—and there were five lakes!"

Miles from any town in their lab on the shore of Lake Huron, Applegate and his crew began testing industrial poisons that arrived daily from factories across the globe. They came labeled with numbers but often with no names; some of the chemical companies submitting their products jealously guarded the formulas in case one turned out to be the lucrative potion that would save the Great Lakes. The program is considered by today's scientists to have been the biological equivalent of a moon shot. But the poison screening system the crew used at the time was anything but rocket science. Workers filled 10-liter jars with water and dropped two juvenile lampreys into each one, along with one rainbow trout and one bluegill. Then they dropped in the poison. The idea was to find a concoction that would destroy the lampreys while leaving the rainbows and bluegills—a suitable proxy for the lakes' native fish species—unharmed.

Cliff Kortman's job was to weigh and mix the powders that arrived from chemical companies around the world.

"All I got was little bottles with a skull and crossbones on it," Kortman told me. He had no college education and was hired initially as a janitor. He said he and the other chemical testers worked with little protection beyond white lab coats and, sometimes, safety goggles or a mask. Kortman remembered how one day he stirred a white powder into a beaker and, as he was walking across the room to put it in the lamprey jar, the solution went poof, literally evaporating into the laboratory air. So he tried it again, and it happened again. Other chemicals were so pungent the room would have to be evacuated. This routine went on for more than two years.

"Imagine testing 40 to 50 unknown chemicals daily," King said. "You just have to keep at it."

Kortman recalled the day he said he opened bottle number 5,209. It was dumped into the jars with the lampreys and the two other fish species, and it didn't take long before the lampreys went limp. The trout and bluegill kept flitting about. "That one was pretty something," Kortman recalled more than a half century later. That one put the scientists on the path to save the Great Lakes.

On July 26, 1957, the *Milwaukee Journal* broke news that the "blind and desperate hunt" for the perfect lamprey poison had succeeded. The first application of it in the wild happened later that year under the cover of darkness, on a tiny creek near Cheboygan, Michigan, with "almost the secrecy of a nuclear project," according to a local newspaper article at the time. Precise dosages of the chemical were pumped into the creek and in the following hours, just as Applegate's crew of "lamprey chokers" had hoped, thousands of the night crawler–sized lamprey surfaced lifeless from the streambed, with no ill effects to any other fish in the area. Applegate described the scene that night as a "real purty sight."

"By midnight, the weary crews returned to Cheboygan for hot lunches. The lid of secrecy was lifted a bit—there were hints, knowing glances," the newspaper reported. "The lamprey had had it."

Further experiments would reveal an even more effective chemical, and by 1961 so much of this "lampricide" was flowing into streams feeding Lake Superior that the lamprey population had been declared under control. Similar poisonings would have a similar impact on the other Great Lakes and by 1967 researchers figured they were well on their way to pushing the Great Lakes' lamprey population down to about 10 percent of its peak, where it remains today due to a nonstop poisoning program that costs about $20 million annually.

But the lamprey solution came too late to save the lake trout in Lakes Michigan and Huron. This was trouble for more than the lake trout and the commercial fishermen who depended on them; at the very moment Lake Michigan's lake trout population was crashing, barely 100 miles to the west, 20th-century naturalist Aldo Leopold was laboring on his seminal work, *A Sand County Almanac,* in which he uncannily captured the critical role an apex predator plays in its eco-system. Writing about mountains that had been stripped of their wolf packs so hunters could enjoy thicker deer herds, Leopold observed:

"I have watched the face of many a newly wolfless mountain, and seen the south-facing slopes wrinkle with a maze of new deer trails. I have seen every edible bush and seedling browsed, first to anaemic desuetude, and then to death. I have seen every edible tree defoliated to the height of a saddlehorn. Such a mountain looks as if someone had given God a new pruning shears, and forbidden Him all other exercise. In the end the starved bones of the hoped-for deer herd, dead of its own too-much, bleach with the bones of the dead sage, or molder under the high-lined junipers," he wrote. "I now suspect that just as a deer herd lives in mortal fear of its wolves, so does a mountain live in mortal fear of its deer."

Had Leopold, who died in 1948, lived long enough to take a day's drive over to the shore of Lake Michigan to witness the bizarre, utterly unpredictable aftermath of the lamprey invasion beginning in the 1950s, he might have found an equally apt way to convey the notion that sometimes a native predator's job isn't merely an essential matter for a functioning ecosystem. It is an existential one.

THE SEA LAMPREY INVASION THAT DECIMATED THE UPPER GREAT Lakes' population of native predators turned out to be just the first wave of ecological trouble unleashed by the 19th-century canal building that opened the Great Lakes to the ocean. After the lampreys slithered their way up the shipping channels came a much more harmless-looking intruder—a type of river herring cherished on the East Coast. Once in the Great Lakes, this fish acquired a new reputation.

Like lampreys, salmon and striped bass, the foot-long river herring normally spend their adult lives in the ocean but spawn in freshwater. Individual female river herring can lay tens of thousands of eggs each spring. Those eggs hatch in a matter of days into wiggly, transparent babies about the size of a grain of rice that feast on flea-sized river plankton with such ferocity that they can grow to three inches in less than two months, and in late summer the now four- or five-inch-long juveniles make their run to the ocean. The fish spend three or four years in the Atlantic before they make the return trip to spawn. It's impossible to peg just how many of these fish swarmed the waters of the East Coast hundreds of years ago, but their range stretched from South Carolina to Newfoundland—and one stream in Maine alone is believed to have carried as many as 100 million juveniles each spring. Only a small percentage of the migrating juveniles survived to return to spawn, but the arrival of these half-pound fish with rich, oily flesh was a time-honored seasonal event. Ancient fire pits littered with fish bones reveal Native Americans have been feasting on river herring for at least 4,000 years and used the fish to fertilize their crops of corn.

The river herring brought out the best of Colonial Americans— they provided a pioneering form of welfare in the 1700s, so plentiful and easily preserved by salting or smoking that they were given away to the elderly and the needy in coastal New England communities. And they also brought out the worst in early Americans—they were shipped to the West Indies for $1 per barrel and fed to slaves.

The fish were important to more than humans. Historical reports show black bears used to scrounge for herring along stream banks. Later, as agriculture progressed in the colonies, pigs snorted their way down to those same creeks to gobble their fill. River herring are also a food source for such bird species as eagles, osprey, great blue herons and loons. In the ocean, they are a source of protein for striped bass, cod, haddock, halibut, blue fish, tuna and even seals, porpoises, dolphins and whales. They eventually became a prime bait fish for the East Coast lobster, crab, cod and haddock commercial fishing industries. But overexploitation and two centuries' worth of migration-blocking dam construction took their toll, and today the river herring run at only a sliver of precolonial levels. In 2011 the Natural Resources Defense Council petitioned to have them listed as a threatened species under the Endangered Species Act, to no avail. Perhaps it is because river herring aren't struggling everywhere.

No one knows how these ocean fish made their way from the Eastern Seaboard into Lake Ontario, where they were first found at the end of the 19th century. Maybe the river herring were actually natives, having made the trip on their own by fighting their way up the St. Lawrence River and into the only Great Lake below Niagara Falls. Some have accused the U.S. Fishery Commission (a predecessor of the U.S. Fish and Wildlife Service) of accidentally planting the species when its biologists dropped a load of similar looking shad into the lake to boost the population of this forage fish to feed its native Atlantic salmon (exterminated by the end of the 19th century) and trout. Or, perhaps, the river herring migrated up the Hudson River, then up the

Erie Canal and then into one of the canal's feeder waterways connected to Lake Ontario.

However the herring got to Lake Ontario, there was little fanfare when the first one was identified in 1873. As was the case for lampreys, the big water of a Great Lake proved to be a suitable replacement for the ocean portion of the herring's lifecycle, and in Lake Ontario the East Coast fish lived for a few years in harmony, if not obscurity, with the native fish. A likely reason is the big lake's predators—Atlantic salmon and lake trout—would have been able to keep them in check. But when commercial overharvests destroyed the populations of those big fish, the now-landlocked river herring began to appear in the 20th century in "almost incredible" numbers, in the not-so-scientific words of Robert Rush Miller, a biologist at the University of Michigan.

It was just a matter of time before they sought new waters to colonize. As with the lamprey, there was a significant lag between the discovery of river herring in Lake Ontario and its migration up the Niagara Falls—bypassing Welland Canal and into the upper Great Lakes. The first river herring found above the falls was in Lake Erie in 1931. They then quickly spread westward, turning up in Lake Huron in 1933, in Lake Michigan in 1949 and in Lake Superior in 1954. Their numbers at first were minuscule; in the early 1950s river herring in the upper Great Lakes were still so rare that specimens were sent to regional museums as a novelty. *Imagine that, an East Coast river herring swimming in our own Great Lakes!*

Had the river herring made their way into the lakes before the lake trout were knocked out, they might have slipped into the food web with scarcely a ripple. But with no predators to keep their numbers in check, they turned viral. Too small to fall prey to the lampreys, the tiny toothless Great Lakes river herring, which are only about half the size of their East Coast cousins, dominated the remaining native fish species by outcompeting them for food and by feasting on their young. If the Great Lakes had been a forest, the lamprey invasion was a fire that burned them down. And the first river herring were the seeds of the weed infestation that blew in afterward.

The result can only be described as an ecological meltdown—particularly on Lake Michigan—previously unmatched in scope or speed. By 1962, biologists estimated the river herring accounted for 17 percent of the fish mass in Lake Michigan. Three years later that number was pegged at 90 percent. The exotic herring had similar success in Lakes Huron and Ontario, and took hold to a lesser degree in the colder and more sterile Lake Superior, and the warmer and more predator-filled Lake Erie.

The alewife, also known as a river herring.

Biologists knew by the mid-1960s the pocket comb–sized fish were swarming in Lake Michigan at extremely high numbers—engine propellers cruising the lake churned the fish up to the surface as if they were bubbles—but nothing prepared them for the size of the dead schools that mysteriously appeared in the summer of 1967, and nobody in the Great Lakes by that point was calling them river herring. They had another word for the hundreds of billions of silvery intruders, and it might as well have been Cockroach-of-the-Inland-Seas, or Locusts of the Lakes. They called them alewives.

THE PILOT FLYING THE NAVY SEAPLANE ACROSS THE DEEP BLUE waters of Lake Michigan must have thought he was hallucinating when he passed over what appeared to be a series of white-as-ice streaks stretching for miles upon miles on the mid-summer lake surface. Or maybe he figured the muggy 90-degree blast of High Plains air pressing down on the lake on this mid-June day in 1967 was kicking back some sort of optical illusion, something known to happen in the Great Lakes. From time to time, residents of Muskegon, Michigan, for example, report seeing the nighttime skyline of Milwaukee, some 80 miles of open water to the west. This, given the earth's curvature, is an optical impossibility. What these people actually glimpse is both real and an illusion. A pocket of hot air sitting above a cool layer can bend the lights of Milwaukee shooting into the night sky eastward, toward the Michigan coast. So the flashing red light of a Milwaukee TV tower that a Muskegonite might see is real, but if he were to set out in a boat to reach it, the baffling light would eventually fade into the black sky.

But the pilot and his passenger from the Federal Water Pollution Control Administration weren't gazing *across* the lake on that breezy day back in 1967. They were looking straight down, surveying the lake's southern end for telltale signs of pollution plumes when they saw a series of white swaths stretching almost the whole distance between the coastal cities of Muskegon and South Haven. The pilot didn't have to guess how large the splotches were; he could calculate them based on the geography of the shoreline. The two cities lie some 50 miles apart. He tipped his wing and dove close enough to the lake surface to know that he wasn't looking at some sort of froth churned up by one of the industrial chemicals so wantonly dumped in southern Lake Michigan during those pre–Clean Water Act days. He had found a mass of belly-up dead and dying fish that numbered in the millions, if not the hundreds of millions, if not the billions. It would have been impossible for him to even hazard a guess, because he had found Mother Nature whacked out in a manner that no freshwater biologist had previously encountered.

The fetid slick of alewives was, mercifully, drifting east, bobbing on the waves toward the relatively unpopulated shoreline of eastern Lake Michigan, where they were destined to rot and eventually wash back out into the open water. But then the winds shifted and pushed the mess back across the lake, toward the 3.5 million residents of Chicago. The first fish carcasses started floating in that weekend. A few days later, 30 miles of Chicago shoreline had been smothered—some places shin-deep—in a mound of rotting fish goo. There had been similar but smaller die-offs across the Great Lakes earlier in the decade, including one the year before that plugged the screens on the cooling water intakes at a Lake Michigan steel plant south of Chicago, causing a loss of a half million dollars a day during a 10-day period.

Yet nothing was like what washed ashore that July, and Chicagoans would never look at their lake the same way. The inland sea that had sustained them for more than 100 years with a marvelous array of native freshwater fillets suddenly started retching millions of pounds of inedible flesh that smelled like human waste. The saltwater native alewives were fantastically good at breeding in the Great Lakes. It just happened that they weren't so good at living in them. In the next several weeks an army of hundreds of workers across the southern end of Lake Michigan used shovels and bulldozers to remove the flesh. Chicago workers reported within the month that they alone had disposed of enough alewives to cover two football fields—500 feet high. But even the city with big shoulders couldn't shovel fast enough. This is how one UPI news report characterized the losing battle: "Chicago was running out of places to bury dead fish, out of money for their removal, and out of people to do that work. A dozen park district employees quit their jobs in olfactory disgust. Morale among those remaining was described as 'low.'"

Those who did stay on the job started cutting corners. "In some cases, the fish are buried to a depth of four or five feet on the beach. Equipment includes sand sifters to separate fish from sand," stated a July 25, 1967, report by the Federal Water Pollution Control Administration. "Although disposal might include burning of dead alewives, such action would result in air pollution. Deodorants have been applied to dead fish and beaches to reduce the stench of decomposition. In some cases, chemicals are used on beaches to control fly maggots in the dead fish."

By the middle of the summer newspapers were estimating that the total cleanup cost on Lake Michigan would reach $50 million—$350 million in today's dollars. The impact spread beyond Chicago. Fish piled up on beaches all across southern Lake Michigan that summer, costing the tourism industry an additional $55 million—again, in 1967 dollars, making this all the better part of what would be a billion dollar problem today.

Not all the dead fish made it ashore; divers scouring the lake bottom reported seeing six-foot-high mounds of carcasses. Yet the lake continued to have an apparently unlimited supply of living alewives. Sonar readings taken around that time showed that a single swimming mass of the fish measured 10 miles long and up to 60 feet wide. The number of fish in that cluster is mind boggling. At the time, a biologist calculated that a mere 15-foot-wide sphere-shaped school of alewives contained as many as 6,000 fish. Some commercial fishermen of the era survived the infestation by figuring out where the shrinking pockets of native whitefish, perch and chubs could be found, but it was almost impossible to avoid running into alewives. Ken Koyen, one of the few remaining commercial fisherman on northern Lake Michigan, can still feel the jolts he and his father suffered when they hit midwater piles of dead alewives while motoring out to their fishing grounds near Washington Island off the tip of Wisconsin's Door Peninsula. "In places they were so thick," he said, "it was like hitting a snowbank."

Some commercial fishermen tried to make a dollar off the alewives by catching them for two pennies per pound, hauling ashore some 40 million pounds of them on Lake Michigan alone that summer of 1967. Food scientists of the era were scrambling to figure out how to turn that flesh into a digestible—if not marketable—form of human food. They explored alewife fish sticks, alewife breakfast sausages and even mixing the alewife flesh into bread dough, molding it into loaves and baking it in industrial ovens. None of that panned out. The only market for the fish was to churn them into cat food, turn them into liquid fertilizer or convert them into fur coats—much of the haul was sold as feed to Midwest mink farms.

By the time the summer die-off ended, estimates of the dead ranged from 6 billion to 20 billion carcasses, each fish dying almost exactly the same way, as described by a biologist with the Great Lakes Fishery Commission: "The stricken fish swam weakly on their sides in vertical spirals that brought them to the surface. Some exerted sudden bursts of swimming effort and were propelled sideways or downward. Their attempts to regain equilibrium lasted several seconds before they again rose to the surface where they quivered and died." But why?

EVEN AS THE 1967 DIE-OFF WAS RAGING, BIOLOGISTS LAUNCHED an exhaustive survey along the "U" that is the southern end of Lake Michigan (from Milwaukee to Chicago to Gary, Indiana to Grand Haven, Michigan) to test the lake for chemical and bacterial trouble. They tested along hundreds of miles of shoreline and waters further offshore for sulfates, nitrogen, chloride, phenol and cyanide. They sniffed for pesticides. None of the chemicals they found were at levels markedly different from where they were before the die-off. Some surmised that the kill was tied to an outbreak of deadly blue-green algae fertilized by sewage spills, a theory later dismissed because similar levels of the algae had been present prior to and after the die-off. Another theory, subsequently debunked, was that alewives were so abundant they had literally suffocated by sucking the oxygen out of the water.

Other researchers cut into the stomachs of alewife carcasses to see what they were eating. The fish kill peaked in late spring and early summer, spawning time and a period when fish typically don't feed, so it would not be surprising to find empty stomachs. Even so, more than half of the fish sampled contained a type of zooplankton that is digested so fast it was clear the fish were finding food—and eating it—right up until the time they died. Others, probably tired of thinking about it, speculated it was just a case of the fish reaching old age. Everyone was left stumped. "The findings did not indicate any extreme or bizarre pollution conditions in the waters that could have caused the massive die-off," concluded the federal report released in the weeks after the first dead fish showed up in Chicago.

The real problem, it turned out, was the alewives themselves. The Great Lakes version of the fish grows only about six skinny inches in length, compared to a fat foot or longer for their ocean cousins. Great Lakes alewives' kidneys are under immense stress because, not being a true freshwater species, the fish are forced to constantly urinate to expel the freshwater persistently seeping into their cells. At the same time, their bodies are working overtime to retain what precious salts they can pull out of the freshwater. Great Lakes alewives also have a stunted thyroid, likely due to a deficit of iodine in freshwater. This may further prime them for death when the real trouble hits: water temperature swings unlike anything the species had to deal with in the ocean. Winds churning deep cold water from the bottom can drop Great Lakes temperatures by as much as 20 degrees in just a matter of minutes.

So by 1967 three of the five Great Lakes—Michigan, Huron and Ontario—were overrun with a rapidly reproducing species ill-suited for living in them. Biologists of the time expected the alewives to rebound at some point due to a lack of bigger fish to eat them, and they expected further die-offs. The lakes were broken, and there was no reason to believe they could right themselves on their own. What had just happened on Lake Michigan proved that.

"To summarize, Lake Michigan was left with a fish population consisting largely of one species, the alewives," stated the federal report in late July 1967. "The natural enemies of the alewives, the predatory fish species, could no longer assert a controlling effect in maintaining a balanced fish population."

Applegate had a plan to solve this problem by restoring the all-but-extinct lake trout populations with a massive hatchery program using eggs and sperm from remnant trout stocks that continued to hang on in Lake Superior. But another biologist had another plan. He didn't want to just resuscitate Mother Nature. He wanted to give her an upgrade by stocking the lakes with an exotic predator that he thought would be a sexier catch than native lake trout.

If Vernon Applegate was the Great Lakes' oncologist who saved them by developing a precision chemotherapy that could be dispensed on an ecosystem scale, Howard Tanner was their plastic surgeon.

Chapter 3

THE WORLD'S GREAT FISHING HOLE

THE INTRODUCTION OF COHO AND CHINOOK SALMON

I n early summer 1968, two businessmen from Waukegan, Illinois, slinked away from work, stopped at a local delicatessen to grab some takeout sandwiches and beers, drove down to the city marina, hopped on a little boat and puttered out into Lake Michigan to go fishing.

It was less than a year since southern Lake Michigan's shoreline had been smothered under millions of rotting alewives. Individual specimens of the invasive fish weigh only about four ounces but, collectively, the little herring had so overwhelmed what was left of the lake's native fish populations that biologists of the time estimated that for every 10 pounds of fish swimming in the lake, 9 pounds were alewives. The businessmen, hopeful they might catch that elusive 10th pound and in a rush because they were on the clock, didn't bother to take off their ties as they dropped their lines in the water about a half mile from shore.

"With sandwiches in one hand, rods in the other, they soon caught several silvery fish that ranged from 3 to 5½ pounds apiece," wrote the *Chicago Tribune's* Tom McNally. "In 45 minutes they caught seven, then they quit and hurried back to work."

Forty miles to the south, on a break wall near Chicago's Navy Pier, an angler that same spring who had hoped only to coax a perch from the alewife-infested waters watched his rod bend and reel whirr as if his hook had snagged the propeller of a racing boat. He eventually muscled in his own 12-pound silvery beast, the likes of which he'd never seen in all his years fishing along the gritty downtown lakeshore. By summer's end "wild-eyed" steel mill workers south of Chicago were hauling in lunkers that topped 20 pounds. These strange fish that attacked a lure with a vigor and viciousness foreign to anything native to the Great Lakes weren't confined to the Illinois shoreline. From Gary, Indiana to Traverse City, Michigan, thousands of fishermen were lured to the Lake Michigan shoreline that summer as word spread of this Great Lake's almost-instant revival.

"The cause of all the furor—and of considerable adventure and misadventure—is a fish new to Lake Michigan, the silver or coho salmon," McNally wrote. "After the introduction of cohos a near miracle occurred in Lake Michigan; the salmon have changed the lake from a near ichthyological desert into an angler's paradise."

A paradise manufactured by man.

It took countless generations for all the various stocks of lake trout to find their place in Lake Michigan and the other Great Lakes. These exotic, hatchery-raised coho salmon, plucked from the Pacific Northwest, bred in hatcheries, and dumped into the Great Lakes by state of Michigan biologists a couple of years earlier, had located their niche in a matter of months, growing from ounces to pounds over the course of just one summer on the flesh of alewives. It took the call of nature for lake trout, alewives and even lamprey to find their own way into the Great Lakes. These Pacific salmon were sent special air delivery over the Rocky Mountains courtesy of the Oregon Fish Commission. It took the Great Lakes thousands of years to reach an ecological fullness that left early European explorers grasping for words to capture its majesty.

These saltwater natives turned the inland seas' reputation almost overnight from the world's greatest lakes, ravaged as they were by the lamprey and alewife invasions, into what one historian termed "the world's greatest fishing hole."

The stocking program that would ultimately pump the Great Lakes full of hundreds of millions of hatchery-raised salmon was one fraught with vast and lasting economic, ecological and political consequences. Before the salmon arrived, the fish in the lakes were managed—to the extent that they were managed—as a publicly treasured resource that put millions of pounds of native species such as lake trout, whitefish, herring and perch on the dinner plates of citizens on both sides of the U.S. and Canadian border—much as a national forest provides lumber for housing construction, or federally owned grasslands are a source of forage for beef producers.

The decision to push aside lamprey-killer Vernon Applegate's goal to restore lake trout and instead focus on grafting an exotic predator onto the Great Lakes was a bit like rehabbing an ailing Great Plains by laying down sod strips of Kentucky bluegrass and turning the place into one giant golf course—one that would require constant tending—rather than reseeding the expanse with native grasses uniquely evolved over thousands of years to provide stability in the face of droughts, fires and roving herds of grazers.

"The salmon weren't put here to feed people," said Great Lakes fishery historian Kristin M. Szylvian, "as much as to amuse them."

The salmon were basically declared off-limits to commercial fishermen and, therefore, off-limits to grocery shoppers or restaurant diners. They became the property of the sportsmen who bought the fishing licenses that funded the salmon-planting program, a program that would prove to be a boon for tourism but also, ultimately, an obstacle in efforts to restore some semblance of natural order to the lakes in the decades after the lamprey infestation.

There are hundreds if not thousands of government bodies—from township boards to state environmental departments to Native American tribes to federal fishery agencies to the U.S. and Canadian International Joint Commission—that have some role in managing the interconnected lakes for the sake of the 50 million people who live within a short drive of their shores. But there was no referendum on this decision to turn them into an angler playground. There was no real public discussion. If there was any debate at all, in fact, it happened in the head of just one man.

BORN IN 1923 TO THE SON OF A GROCER IN NORTHERN MICHIGAN, Howard Tanner started fishing on Sunday mornings with his father at age 5. The two chased brook trout near the railroad tracks along the Jordan River in Antrim County—the same region fished by a young Ernest Hemingway several years earlier. Tanner remembers the catch limit at the time was 15 fish per day and, typical of the era, the Tanners were not catch-and-release guys. Especially after the country entered the Great Depression and his father lost the store, became sheriff and moved the family into the living quarters of the county jail.

"My grandmother and my mother's younger brothers and sisters lived just down the street, and so we ate all we caught. There was no question about that," the 91-year-old, barrel-chested Tanner told me at his ranch-style home outside of Lansing. It was a surprisingly modest one for a man whose career was decidedly not.

The young Tanner became so skilled at putting fillets on family plates that by age 15 he had printed up business cards declaring himself a professional fishing guide and was taking city anglers to inland lakes to chase smallmouth bass or for fly fishing on the Jordan River. The fact that these men fished just for fun—that the whole enterprise of dropping a line in the water could have nothing to do with what would be on the dinner table that night—made an indelible mark on the teenager.

"Sometimes I say it was the best job I ever had—I got paid to go fishing," he said. "I met quality people. They were mainly wealthy men, I figured, who paid me $2 a day to take them fishing."

After high school Tanner was pulled away from the guide business to get an education. "I turned 18 and started college at Western Michigan University, where my mother and her sisters and brothers had all gone to school," Tanner, who described himself as an unfocused scholar at that point, told me with the same sort of pathos that you might expect from a one-time baseball prospect as he looked back at the path that took him away from the thing he loved most.

"And then in December the Japanese dropped the bombs on Pearl Harbor," Tanner said as a wry smile slowly cracked on his face. "That sort of disrupted things."

By age 21, Tanner was a soldier helping to carve airstrips out of the jungles of the South Pacific. He was also in the throes of a long-distance courtship with a girl named Helen whom he'd met during his stint at college in Kalamazoo. She sent him letters every day. Some were more business than love. "Helen is very organized," Tanner said, "and she had words to the effect of—If I'm going to marry you, what are you going to do?" Tanner had some brief training early on in the military as an engineer, but that did not go so well; he is quick to point out that he did not relish the classroom as a young man. But he was a hell of a fisherman. When Helen pressed, he told her he thought he might be able to make a living as a biologist. More letters followed. "She sent me the curriculum from about six universities that had fishery programs, and Michigan State University was one of them."

By age 29 Tanner had returned from the war and acquired a knack for studying, a doctorate in fisheries biology from Michigan State University, and a wife named Helen. Tanner's first job after graduation was as auspicious as it was ambitious. He turned life upside down in a lake in the middle of a Michigan state forest using a generator, a pump and a pipe to suck the water 40 feet up from the lakebed to the surface. "The research reason for that was the nutrients would gradually settle to the bottom of the lake, but there was no oxygen down there for biological activity and the question was: What would happen if you pumped that nutrient-rich water back up on the surface where there was sunlight and life?"

As the junior scientist in the experiment, it was Tanner's job to maintain the generator that ran around the clock on the shore of West Lost Lake. He vividly remembers one early summer morning in 1952.

"There was a fisherman sitting on the bank with his rod, smoking a pipe as I put gas in and checked the oil, and I went over to say good morning. He said: 'Could you tell me what you're doing?' And I said: 'Yes, we're sucking the water off the bottom of the lake and putting it up on the top.' He looked at me and said: 'That's just exactly what I thought,' and walked away," Tanner recalled with such a hard, hoarse laugh he had trouble getting the words out. "I can hear him in the bar saying, 'You know what I saw today?'"

The experiment did what was expected—it sparked a bloom of plankton near the lake surface. But the flicker of life flared out once the scientists turned off the pump. "It probably resumed its normal situation in a week or two," Tanner said with a shrug. He did not stick around long enough to find out.

At summer's end Tanner headed in a direction he did not expect—westward with Helen and their two young sons to Colorado, where he had landed a job at what is now Colorado State University. Since Michigan's state conservation department paid for his education, Tanner had expected the department would hire him. He was disappointed at the time. But looking back more than six decades later, he said that move was exactly what was needed for his own professional development—and for the future of the Great Lakes.

Tanner found the Western approach to fishery management "totally different, in almost every way" from how biologists approached the job in the Great Lakes region. The West has some of the world's most renowned trout rivers and streams, places like Montana's Blackfoot River, on which writer Norman Maclean learned to fly fish and later immortalized in his book *A River Runs through It*. There is Silver Creek in Idaho, a spring-fed stream south of Sun Valley that prompted Hemingway, upon floating it for the first time in the 1930s, to gush in a letter to his son that from its crystalline waters he witnessed "more big trout rising" than anywhere he'd ever been. And there are the pristine rivers of the national parks, including the waters of Yellowstone, home to the largest remaining inland population of cutthroat trout—an icon of the Wild West.

But many of the water bodies popping out of the West's dusty landscape are artificial pools backed up behind concrete or earthen dams. These were places where fishery managers were not worried about maintaining or propping up stocks of native fish by restricting harvests or resuscitating them with hatchery programs. These manmade water bodies were, essentially, a blank canvas for a biologist to construct an ecosystem from scratch. "When you create new water and there is nothing in it," Tanner explained, "you plant something."

Hatchery-raised sport fish like trout and bass were planted with abandon in the reservoirs and hatchery-produced eggs were shared in a manner that created bonds between fishery managers and researchers stretching across state lines. This build-it-yourself approach to ecosystem management, and the friends Tanner made in top state fishery jobs across the West, proved immensely important not long after a phone call came in 1964 from one of Tanner's old professors who wanted to know how his former student's career was progressing.

"I was completely settled in. We'd been there 12 years and my wife had her activities and things she was doing with the community and the schools. We had three sons by this time and they were all in their programs at school. And I was happy," Tanner recalled, noting that he had recently been promoted to Colorado's chief of fisheries research. The professor pressed.

"Did you know that Michigan is looking for a chief of fisheries?" he asked. Tanner said no. "Well," said his professor, "would you apply?" Tanner said he would think about it.

He had personal reasons to return. His father was battling cancer and his father-in-law also had been ill. There were also professional allures, not the least of which was the vastness of the waters lapping at the borders of Michigan, the most coastal of the lower 48 states, and one with a most oddly nautical motto: *Si quaeris peninsulam amoenam circumspice*—If you seek a pleasant peninsula, look about you.

Tanner thought about a long-ago ferry trip across Lake Michigan that took him into waters so endless he might as well have been in the middle of the ocean. He thought about the blue expanse he once fished off Beaver Island. The numbers started to roll in his head, and they weren't just the extra dollars the state of Michigan was willing to pay him. Tanner might have become a big fish in Colorado, but he knew he was working with mere ponds compared to the big waters back in Michigan: "About 50 percent of the surface freshwater in the 50 states are within the boundaries of Michigan, and the other 49 states shared the rest of it," Tanner said. "It was a big job."

And it was about to get bigger. For decades the federal government had taken a lead role with the Great Lakes states in managing the open waters of the lakes for commercial fish harvests. When the fish stocks crashed in the early 1950s, the U.S. and Canadian governments got together to create the Great Lakes Fishery Commission to combat the lamprey infestation and coordinate fisheries management over what was left of the Great Lakes native species. As Tanner recalls, the original idea was to give the bi-national commission authority over Great Lakes fisheries in a manner that would have jeopardized the individual states' right to act independently within their own waters. He said one state in particular balked at ceding its authority to the commission—"Blessed Ohio."

Blessed because, in the end, he said it was decided that the Great Lakes Fishery Commission could advise but not dictate. And because, by the early 1960s, after decades of deferring to federal fishery biologists who had long focused on Great Lakes native species like lake trout and whitefish, the state of Michigan was ready to take the lakes in a different direction—away from a focus on commercial harvests and toward a future where fun trumped natural function.

Tanner took the job, and his new boss gave him one simple directive: "Get out there and do something spectacular."

BY THE TIME TANNER RETURNED TO MICHIGAN IN THE FALL OF 1964, Applegate's lampricide was already coursing down the lakes' tributaries and the lampreys were well on their way to being knocked down to about 10 percent of their peak. But this did nothing to solve the alewife infestation. On one of his first days on the job, Tanner grasped the scope of the spreading scourge while on an aerial tour of Lake Michigan. As the pilot cruised over an alabaster blotch just off Beaver Island, Tanner asked him to drop lower and take another pass. This was three years before the alewife explosion that would swallow southern Lake Michigan's shoreline, but massive slicks of dead fish like the one below the twin-engine plane that day were already becoming common. Tanner asked how big of a mess he was looking at. The pilot told him it was about seven miles long and two-thirds of a mile across. The surface area of that single slick of floating flesh was about the size of the largest lake in Colorado. "That was the first eyeball experience I had," said Tanner. "That was a very, very impressive sight."

Although most at that point considered the ever-ballooning alewife population a natural disaster, Tanner assessed it as an opportunity to raise bumper crops of predator fish. The natural choice, and the one pursued at the time by the newly formed fishery commission, was to bring back the beleaguered lake trout, small stocks of which continued to hang on in parts of Lake Superior and northern Lake Huron. The problem for the lake trout was that it was not favored by the new Michigan fishery boss. Tanner had inherited an essentially blank slate— almost like a freshly filled reservoir in the West—and he saw it as a prime opportunity to give Mother Nature an upgrade.

Tanner and his colleagues in the Michigan Department of Conservation saw the state's Great Lakes waters as the Midwest's last frontier for recreation, one that could generate immense economic activity because of the time, boats and gear it would take for sportsmen to chase fish on such big water. Converting the Great Lakes fishery from one that was essentially managed as a self-sustaining public food supply into one that was intensively managed for the thrill of recreational anglers was, for Tanner, a natural evolution. "You manage the resource to produce the greatest good for the greatest number for the longest period of time," Tanner said, borrowing the axiom of the first boss of the U.S. Forest Service, Gifford Pinchot, who viewed forestry as "the art of producing from the forest whatever it can field for the service of man."

"And for a century, probably, commercial fishing fit that criteria," he said. "But in 1964 it was long past. Our inland lakes were crowded with water skiers and our trout streams were crowded with canoes and there was a fair amount of mobility and expendable income."

Tanner concluded that lake trout just weren't sexy enough to draw anglers out onto the treacherous Great Lakes. His knock on lake trout is they did not put up much of a fight when hooked. It's not for lack of heart, but partly because of an inability for the fish to quickly expel the gas in their "swim bladder" that allows them to adjust buoyancy so they can hunt smaller fish at greatly varying depths. The rapid loss of pressure as one is reeled in from the deep can inflate that bladder to the point that the fish pops to the surface almost like a balloon, mustering all the fight and wriggle of a snagged rubber boot.

"I've never been against lake trout," Tanner said. "I'm just not very enthusiastic about lake trout. My own experience is lake trout aren't a lot of fun to catch."

One of his first decisions as fishery chief was to send his lieutenant, Wayne Tody, in search of a new predator fish. The initial idea was to restock the Great Lakes with striped bass, a jumbo predator that can grow to more than 100 pounds and roamed along almost the entire Eastern Seaboard. Fishery workers in South Carolina had already mastered raising "stripers" in hatcheries, and previous stocking enterprises in Western reservoirs showed the fish could spend their whole lives in freshwater lakes by devouring a little schooling fish called the gizzard shad, which somewhat resembles the alewife. Everything Tody saw on a trip he took to South Carolina convinced him that Lake Michigan and the other Great Lakes would be a perfect fit for the Atlantic bass. The food source was obviously there and existing stocking programs on inland waters had already answered the question of whether the fish could thrive exclusively in freshwater. And, most importantly, stripers grow fast and fight hard when hooked.

But bad news came after a South Carolina bass specialist whom Tody brought back to Michigan found the lakes lacked the long, slow-flowing, undammed rivers that the bass need to spawn. He also determined the lakes were too frigid for the stripers to grow fast and big—a prerequisite in Tanner's quest to create a sport fishery.

Tody then took a trip to the West Coast to learn about Pacific salmon, a species built to feast on schooling prey fish, fish just like alewives. Salmon thrive in waters as cold as the Great Lakes. The waters of Lake Ontario below Niagara Falls once, in fact, teemed with Atlantic salmon before they were exterminated in the 19th century by overfishing and habitat destruction. Perhaps most importantly, salmon can expel their swim bladder gas fast as a burp in a manner that allows them to take their ferocious fight all the way to the deck of a boat.

There was the question of whether a Pacific Ocean native could spend its entire lifecycle in freshwater. Some Canadian biologists whom Tody talked to during the trip scoffed at the idea. But he found other biologists who thought a Great Lakes stocking program might work with coho, a smaller species of salmon that spends about half of its three-year lifecycle in freshwater before it swims to the ocean to spend about 18 months—two summers and a winter—chasing Pacific prey and then returns to its native freshwater streams to spawn and die. There were even a couple of examples of successful salmon stocking experiments in freshwaters of the West. But there were also plenty of examples of planted salmon struggling in freshwater, including dozens of smaller-scale attempts in the Great Lakes that dated back to the 1870s. Still, nobody had tried planting salmon in the alewife era or, as Tanner put it: "Nobody ever had that much water with that much food supply."

Tanner had learned enough from Tody's Western research expedition to conclude Pacific salmon would be the future of the Great Lakes, even though he was not sure the transplanted fish would reproduce naturally. Although he thought it possible, he was not banking on it and was prepared to embark on an annual stocking program that could last years, decades, even longer. He just did not know how to get it started. He had tried for years back in Colorado to acquire coho salmon eggs from colleagues in the Pacific Northwest to try and plant in Rocky Mountain reservoirs. For years he had been rebuffed because Northwest hatchery workers, trying to bolster their own wild stocks, were having a hard time figuring out how to raise the fish in captivity and had no eggs to spare.

The development of a fish food called the "Oregon Moist Pellet" changed that. Prior to the early 1960s, hatchery workers raising salmon had to grind things like salmon eggs, livers and spleens daily to feed their baby fish. It was an intensive, hit-or-miss process that hampered production of viable fish crops. But the invention of the vitamin-dosed, pasteurized pellet made of things like meal from cottonseed and wheat germ as well as the guts of salmon and herring could be whipped up in industrial-scale batches, frozen and dispensed daily. It led to a boom in raising salmon out West and that, eventually, led to a phone call Tanner got one night during his first fall back in Michigan. An old colleague said that Oregon might have some coho eggs to share.

"I didn't believe it," Tanner said. "I knew it wasn't true, because I'd asked for coho eggs for years. And they didn't have a surplus."

Tanner was so excited that night he couldn't go to bed. He just sat in a chair, the lights on—in his living room, and in his head. "I'm thinking if that was true, then the opportunity was there. It was—it just was crystal clear. I mean, everything would fit. There would be a food supply. The waters were suitable in temperature. . . . And if I chose to do it, we could do it."

Tanner made a call to Oregon the next day and found biologists really did have salmon eggs to share, due largely to the newly concocted fish food. "It's the Oregon Moist Pellet," the biologist on the other end of the line told him. "These fish are coming back to spawn. They're more vigorous. They endured."

In December 1964—less than four months after he took the job— the first batch of an initial gift from Oregon of one million coho eggs was loaded on a plane bound for Michigan.

THE IDEA OF PURPOSEFULLY INTRODUCING AN EXOTIC SPECIES
into the Great Lakes had been controversial for decades before Tanner
made that call. In the 1870s, the U.S. Fish Commission, a predecessor
to the U.S. Fish and Wildlife Service, launched a national hatchery and
stocking program for common carp, which early European settlers val-
ued as a food fish. The program was so high-profile that the agency
chief built carp-rearing ponds on the grounds of the Washington Mon-
ument after the Civil War. By the early 1890s the bottom-grubbers that
wreak ecological havoc by ripping up vegetation that stabilizes lake bottoms were muddying Great Lakes waters. Then came the introduction
of finger-sized rainbow smelt, an Atlantic Ocean native initially stocked
in an inland Michigan lake in 1912. The fish soon escaped into Lake
Michigan and then the rest of the Great Lakes and ravaged the natural order of things by gorging upon the young of native species like
lake herring, whitefish, perch and lake trout. These two intentionally
sparked scourges led to an admonition from the U.S. Department of
Commerce in a 1926 report arguing against any future exotic species
stocking programs for the Great Lakes.

"We have already accumulated so much experience from the introduction of foreign species of vertebrates that it would seem unneces-
sary to caution against a continuation of the practice," wrote a biologist
from the U.S. Bureau of Fisheries, "and it is to be hoped that no organization will in the future assume the responsibility of the importation
of any uncontrollable indigenous animal."

But nearly 40 years later Tanner was more than willing to accept
such a responsibility, explaining that he and his colleagues attacked the
job of remaking the Great Lakes with an intensity that reflected their
collective war experiences. Tanner had built airstrips in jungles of the
South Pacific as a member of the Army's Signal Corps. His boss had

led Marines ashore. And that guy's boss had been a bomber pilot over Germany. "If something needed to be done," Tanner said of the mind-set of the battle-hardened group, "you did it."

"They didn't consult Canada. They didn't consult the other states," said Great Lakes fishery historian Szylvian. "They just did it."

The salmon plan did indeed have regional and international ram-ifications, because the fish that swim thousands of miles in the ocean could theoretically roam across all the lakes, oblivious to the lines on the map that separate the waters of Minnesota, Wisconsin, Illinois, Indiana, Ohio, Pennsylvania, New York and Ontario. The salmon plan also put Michigan on a collision course with federal biologists who had already begun trying to restore native lake trout with hatchery plant-ings. Although Tanner and Tody publicly said they hoped commercial fishing could continue at some level and that they hoped lake trout would make a recovery, they made it clear this was not their priority. They did not want to try to turn the clocks back on the lakes to recap-ture some of their natural bounty and balance. They wanted to turn them into a 20th-century angler playground.

"The coho is aimed at a specific fisheries management problem—namely to elevate the fisheries resource of the Great Lakes to its maximum potential for recreational fishing," Tanner and Tody acknowl-edged in a 1966 report they released just weeks before the first class of hatchery-raised salmon were to be planted.

Tanner's initial plan for "farming" the lakes was to stock the salmon for three consecutive years to see if any returned to the streams in which they were planted at the end of their three-year life. The idea was, at the end of the third year, the returning fish could be captured and their eggs and sperm harvested to maintain a hatchery breeding program. But Tanner knew it was also possible Mother Nature might do that job herself.

"It is certainly possible," he and Tody wrote in their 1966 report, "that success may be achieved in establishing the coho in just the first life cycle. At this point only an actual Great Lakes introduction can provide additional information."

Just like the little experiment on West Lost Lake 14 years earlier, Tanner had again set out to turn life upside down in a lake. But now the scope of his ambitions had reached a Great Lakes scale.

Sixty years later, Tanner still cannot believe he was allowed to do it. "All my life," he said, "I have marveled that one person, that happened to be me, was given the opportunity and the authority to make a decision of this magnitude."

On a gray, snowy April 2, 1966, Tanner, wearing a tie and overcoat instead of the cowboy hat he was fond of wearing in the field, took a microphone on a makeshift stage on the banks of the Platte River flowing into Lake Michigan southwest of Traverse City. Dignitaries sat on card table chairs behind him for a brief ceremony before a state legislator picked up a ceremonial golden bucket and dumped a load of finger-sized cohos into the river.

Later that day, Tanner tipped his own bucket of coho into a nearby creek—one of the very creeks where Hemingway had fished for trout a half century earlier. It was a bittersweet moment; Tanner had already quietly agreed to take a new job as a professor at Michigan State University.

"I stood at the banks of Bear Creek," he told me, "and the truck left and the photographers left and I stood there in the snow watching those fish go down to the main stream wondering—how soon and how big?"

Would the fish survive to adulthood, and if so would they return to the rivers and streams in which they were planted? Would they, as some scoffed, swim east instead for the salty allure of the Atlantic Ocean? Would they just become fish food? Or would they begin reproducing on their own and alter life in the Great Lakes in a manner no one could predict? Tanner took his worries home and confessed them to his wife as they sipped cocktails.

"I remember telling her, I'm going to be either a hero or a bum," he said. "Whichever it is, it's going to be loud and clear. And it's going to reverberate for a long time."

ANY DOUBTS TANNER HAD ABOUT WHETHER A HATCHERY-RAISED crop of Pacific salmon could find a home in the freshwater of the Great Lakes evaporated just months later. The fish began their lives in a Michigan hatchery in the winter of 1965 and were raised by human hands until their release in April 1966. Cohos typically have a three-year life cycle. They spend the first year and a half in the rivers and streams in which they were hatched before descending to the ocean for about 18 months. Then they return to their native streams in the fall of their third year to spawn and die.

This meant the first crop Tanner planted would have been expected to spend the summers of 1966 and 1967 in the lake until, with the arrival of the second autumn, it was hoped that they would follow their exquisite sense of smell back to the waters into which they had been released. But not all Pacific salmon follow this three-year plan in their native saltwater. Sometimes, if conditions are right, a handful of cohos will grow fast enough to mature sexually and return to spawn after just one summer in open water.

This is exactly what happened on Lake Michigan in fall 1966. Fishermen pulled hundreds of coho from waters near where they had been planted just several months earlier. Some were already a whopping seven pounds, approaching full size for a typical adult coho in its native Pacific waters. More remarkably, in the ocean these early returners, referred to as "jacks," are almost always males. But Michigan's first jack class included females—a further sign that Lake Michigan was fine salmon habitat indeed.

Tody, who took over the fishery department from Tanner after he left, offered fishery officials from the Northwest an all-expense-paid trip to the Great Lakes to thank them for helping to launch the most ambitious freshwater fish-stocking program the world had ever known. The delegation arrived in Houghton near the northern tip of Michigan's Upper Peninsula in September 1966. The group spent the next three days on an aerial tour of the upper Great Lakes, flying first to the eastern end of Lake Superior then down the western Michigan coast, over to Chicago and then back north, over the Straits of Mackinac and down the eastern Michigan coast of Lake Huron. The Washington state chief of fisheries was left flabbergasted by the scope of what he'd only known as blue blobs on a map, remarking that "never had he imagined that much water could exist outside the ocean," the now-deceased Tody recalled in his history of the salmon program.

Tanner, who participated in the festivities, remembers the Westerners' optimism after learning about the huge number of full-grown coho jacks that had returned just months after they were planted. The visiting officials told them that jack run was nothing compared to what they should expect the next year, when the majority of the spawners from the nearly one million planted coho were expected to return to the waters in which they were released.

"This is going to be big," they told him. "It's going to be big."

JUST AS THE ALEWIFE DIE-OFFS ON THE SOUTHWESTERN SHORE of Lake Michigan started to wane in the fall of 1967, some 300 miles away on the northeastern shore of the lake came another fish run— the first full class of coho returning by the tens of thousands to their "native" waters. The salmon, feasting just as expected on the hordes of alewives, were coming in fat like footballs, some weighing more than 20 pounds; an adult ocean coho's average weight is about 8 pounds. Tanner told me it's hard to understand the excitement this generated among the thousands of sportsmen who descended on the lakeshore that fall. It was as if all the skiers in Michigan awoke one morning to find that their little hills had been replaced by the Rocky Mountains. "Try to imagine a population of avid fishermen . . . they might have dreamed of going salmon fishing to the West Coast and spending a lot of money. Or maybe go to Alaska or something like that but very few of them would ever do that," said Tanner.

"And suddenly with existing tackle and small boats and motors, they went out and they caught their load of five fish, and their lines got busted, and the fish were leaping out of the water, and they were all around them and the excitement was just explosive. It was a frenzy."

Towns in northwestern Michigan were swamped in a manner that nobody had planned for. "Motels filled for 50 miles around, parking lots were hastily constructed, fishermen sometimes waited in a mile-long line to launch a boat," said the Michigan Department of Natural Resources in its 1967–1968 Coho Salmon Status Report, "restaurants ran out of food, tackle dealers ran out of tackle, and gas stations ran out of gas."

Near riots erupted along the streams and riverbanks up which the salmon were swimming, and at one point more than 60 conservation officers backed by state patrols had to quell what one officer called a "mob attack" on the fish. "The one thing we did not fully anticipate was the fever, or 'salmon craze,' that fishing this new exotic species generated," Tody confessed to the Associated Press at the time. "Anglers form lines across the streams and methods of taking the fish include snagging, spearing, the use of dip nets and clubs and even grabbing the fish with their bare hands."

Things turned deadly out on the big water during the peak of the craze on September 23, 1967, when thousands of fishermen along the northwestern Michigan coast ignored—or did not understand the significance of—the red flags snapping in the wind that signaled a small craft advisory. Some pressed miles out into the lake, some were in boats as small as inflatable rafts and canoes. A gale blew in just as predicted. Hundreds of boats were swamped and U.S. Coast Guard helicopter crews plucked dozens of people from the water. Not everyone made it off the lake alive. News of the disaster spread as far as the West Coast where headlines the next day said dozens of fishermen were missing. The actual death tally ended at 7. "All bodies eventually washed ashore," reported the Associated Press. "Not one was wearing a life jacket."

The deadly day only prompted fishermen to invest in bigger boats and engines; just months later, Wisconsin-based motor builder Evinrude began marketing special salmon boats. "This is Coho country," proclaimed a display ad in the May 17, 1968 edition of the *Oshkosh Northwestern*. "The more you leave to Evinrude—the less you leave to luck." An Evinrude spokesman at the Chicago National Boat, Travel and Outdoors show that spring walked the convention floor, according to the *Chicago Tribune*, with a "twinkle" in his eye, proclaiming that the introduction of cohos to Lake Michigan would trigger one of the greatest spikes in outboard motor sales in industry history.

"Never before in fishery management has such a result been achieved," he said. "Its implications are awe inspiring."

Local communities benefited as well; retail sales in the northeastern Michigan region most affected by the salmon craze jumped by $11.9 million during the three-month salmon run of 1967 compared to the same period the year before. At the same time more boat ramps, marinas and motels were hastily being constructed. It was the exact coastal economic boom Tanner had hoped for, and it soon spread across all the Great Lakes.

In 1967 Michigan expanded the stocking program to include chinook salmon, similar to coho but substantially bigger. Chinook, also known as "kings," can grow to more than 100 pounds in their native Pacific waters and would top out in the Great Lakes at more than 40 pounds. But chinook aren't just bigger. They are much cheaper to raise because they can be hatched, reared and planted in six months rather than the 18 months it takes to launch a class of coho. So by the fall of 1967 Lake Michigan was bursting with nearly two million planted coho, one million chinook, and about one million native lake trout, which had been planted by federal biologists more focused on returning some semblance of natural order to the lake than throwing fuel on the salmon explosion. The other Great Lakes states immediately developed their own salmon stocking programs with free eggs provided by the state of Michigan. By 1968 stocked Pacific salmon were swimming in all five of the Great Lakes.

"There was so much pressure for us to follow," said Lee Kernen, retired fisheries chief for the Wisconsin Department of Natural Resources. In 1968 he said Michigan biologists gave him 25,000 hatchery-raised salmon to plant on the Wisconsin side of Lake Michigan. Kernen did the work himself, and he could not believe what he saw the moment the fish hit the water. "They'd be swimming around with big alewives sticking out of their mouths like a cigar," he said. "You knew it was a can't-miss, no-brainer. They were going to grow big, and people would have the best fishing they'd ever seen. And they did."

The few Lake Michigan commercial fishermen who remained in operation at the time were prohibited from harvesting salmon or lake trout, so some began casting their nets for alewives, which they were selling as pet food and fertilizer for a penny or two per pound. Crews were already hoisting tens of millions of pounds of alewives out of Lake Michigan by the mid-1960s and federal biologists wanted to ramp up the harvest. Reducing alewife numbers would not only help the commercial fishermen but also provide a boost for the lake's native species whose eggs and young were being decimated by the alewife hordes. Tody recalled the regional U.S. Bureau of Commercial Fisheries' office had "demanded" authorization from the state of Michigan to build fish meal plants to expand the alewife harvest. Worse, Tody said, he had to endure the federal agency's criticism of the salmon stocking program as a gimmicky "flash in the pan" fishery experiment.

"This," Tody recalled, "was more than we could stand."

He took his concerns to the Michigan governor, who took them to Washington, D.C. The result: although the federal lake trout stocking program would continue at a far smaller scale than the state salmon program, the Bureau of Commercial Fisheries regional office was shuttered in Michigan, and its staff was transferred to Gloucester, Massachusetts.

"It was no longer an asset to us," Tody concluded. "In fact, it was working at cross-purposes with our entire program"—a program entirely reliant on the invasive alewife that Tody considered as important to the future of the Great Lakes as any of its native species, perhaps more so.

"Someday," Tody quipped to a reporter in 1968, just a year after Chicago's beaches were smothered in rotting alewife flesh, "we may be raising alewives as food for our sport fish."

ALTHOUGH TANNER TODAY TAKES PARTICULAR PRIDE IN THE economic transformation he launched, he says the legacy of the program extends beyond a boost to boat builders, charter fishing operators and the economies of coastal communities. After the salmon arrived, he notes, the public demanded action to ensure the fish they caught were safer to eat. The Great Lakes at the time had been subjected to more than a century of industrial, municipal and agricultural pollution and Tanner cites the salmon as a big reason people demanded more be done to protect the lakes. The pressure to do so was immense, especially when biologists started digging into the salmon flesh and learned it was packed with the pesticide DDT in concentrations as high as 19 parts per million, more than three times the federal limit of 5 parts per million. A budding program to commercially sell excess salmon captured in Michigan streams was scrapped but sportsmen were free to continue to catch their fill; the Associated Press reported at the time that the "scare" of the high DDT concentrations was of little concern to Tody, who noted it "hasn't bothered the real fishermen at all."

"We find that cooking them gets rid of most of the oil—which contains the DDT—and gets them down well below the five parts per million," he said. "And they taste damn good."

Yet the salmon contamination problem prompted Michigan to lead the way in banning DDT in 1969, and although fish consumption limits continue to this day for PCBs and other contaminants, Great Lakes waters are by most measures far cleaner in terms of industrial and municipal pollutants than when Tanner arrived on the job in the mid-1960s. "I always point out that we created a constituency for the lakes," said Tanner, who was named Conservationist of the Year in 1968 by the National Wildlife Federation. "There was an awakening."

But to this day Tanner bristles at the idea that he was brought back from Colorado to solve an environmental problem—to figure out how to control alewives. He said their numbers just happened to plummet the year after coho fever hit in 1967.

"An introduced species such as the alewife characteristically builds up to a peak and then the population crashes. And in the early summer of 1967, the alewife population crashed, and there were dead alewife like nobody had ever seen before," he said. "It was the worst."

Then the first run of coho hit later that fall, drawing thousands of anglers to water and making headlines in national newspapers and magazines. And the next year, there were no dead alewives on the beaches. "Everybody *knew* that the salmon had eaten those alewives," said Tanner. "Well, there wasn't enough salmon in Lake Michigan that year to make a dent in the population. And we said that. We said 'No, no, no, the salmon didn't do it.' We said that for at least two months"— all to no avail—"and then we said: 'Okay, we did it.'"

It was good publicity for the Great Lakes' salmon stocking program, which eventually became dominated by chinook plantings. But it perpetuated a myth Tanner said is "degrading" to everyone involved in the project. "We were fisheries biologists," he said. "We were not there to solve a beach problem. We were there to build a fishery." He likened his philosophy to that of a rancher who stumbles upon an island the size of Lake Michigan that is overgrown with grass. "Do you think he'd say, 'I could put some cows on that island and shorten that grass?'" he asked. "That is not what he's going to say. He's going to say, 'My God, I can raise more beef than you ever saw in your life.'"

Just like ranchers trying to squeeze every possible ounce of beef from a pasture, biologists boosted annual hatchery plantings into the millions in each of the Great Lakes in the 1970s and 1980s. No lake was more heavily "farmed" in this manner than Lake Michigan, which by the mid-1980s was being pumped with more than 10 million salmon annually while federal lake trout plantings were only about half that.

"Tanner got his dream," said historian Szylvian. "They were all on TV with those big old fighting salmon. And everybody thought: Oh my God, we can go to Chicago, rent a boat and be right in view of the skyscrapers catching these kick-ass salmon!"

By 1985 Tanner was busy developing a super strain of salmon he hoped would grow to twice the size of the lake's trophy chinook. The idea was to heat-shock fertilized chinook eggs in a manner that would give the fish three sets of chromosomes instead of two. The end product would be an adult fish that would never reach sexual maturity, which meant none of its energy had to be wasted on sperm or egg production.

Everything would go into growth, and because its biological clock was frozen, the fish would not be shackled to the salmon's three-year life-cycle, enabling it to grow for up to a decade or longer.

"They are trying to decide how many of these fish should be stocked, and how the forage base will be impacted," *Field & Stream* wrote the year before the program started. "Since baitfish populations in Lake Michigan are currently down, the scientists are paying close attention to this factor."

They should have been paying closer attention. By 1988, nearly a million of Tanner's "super salmon" were swimming in lake waters or growing in fish hatcheries, and Tanner was tantalizing reporters with hints that 125-pound chinook could be on Lake Michigan's horizon.

But it turned out, like cows overgrazing a pasture, there was a limit to the number of salmon the lakes could sustain. Rotting chinook started washing ashore on Lake Michigan later that same year, dead from a bacterial kidney disease at least partially induced by starvation due to the lake's steady decline in alewives, the population of which by then was only about one-fifth of its 1960s peak. The monster-salmon program was scratched, but even so the chinook catch rate by the early 1990s plunged to just 15 percent of what it was in the mid-80s.

The salmon famine only further enhanced the alewife's image among sportsmen as a treasured forage fish that was the backbone of the Great Lakes' multibillion dollar fishery, even if it was breaking the back of the native species that were left; there has long been a known correlation between suppressed numbers of egg-gobbling alewives and increased numbers of native species. Yet since the dawn of the salmon program Great Lakes states showed little interest in suppressing ale-wife numbers in the name of boosting native fish. In fact, when the alewife collapsed on Lake Michigan in the early 1990s Wisconsin biolo-gists declared them a species in need of "protection" and halted the last remaining commercial alewife harvests. The protection order worked as expected; alewife numbers rebounded but native perch, which had been recovering with the alewives' demise, crashed.

This was the moment when many commercial fishermen whose livelihoods had been ravaged by the decline of native species believe state biologists went too far in farming the lake for salmon. Wisconsin commercial fisherman Pete LeClair saw commercially harvesting alewives as akin to tending weeds; suppressing their numbers was the best way to protect the fry, or young, of native species like perch which, along with whitefish and chubs, were basically all that was left of Lake Michigan's commercial stocks at the time.

"Don't tell me the alewife don't eat the goddamn perch fry," LeClair howled in an interview more than a decade after the alewife fishery was closed. "When the alewife come in, it's just like a herd of cattle." Wisconsin banned commercial fishing for perch on Lake Michigan in the mid-1990s to try to resuscitate that species. But it has never recovered.

"It was a good idea, but they overdid it," LeClair said of the salmon. "Now they just plant, plant, plant. They have no idea what they're doing out there."

By the late 1990s more than a half billion hatchery-raised Pacific salmon had been planted in all the Great Lakes in the years since Tanner dumped his first bucket in 1966. And then came a real crash.

THE AUTHOR OF THE BUMPER STICKER MAXIM THAT "A BAD DAY Fishing is Better than a Good Day Working" never sat on the shore of Lake Huron with Jay Hall during the fall salmon run. The 47-year-old mechanic from Flint was perched on a collapsible chair on a windless fall day in October 2014, hands stuffed in his jacket pockets, fishing pole propped in the crotch of a twig stuck in the lakeshore muck. His gear would have been dragged out to sea had a salmon hit, but Hall wasn't much worried about that. He was at the end of a bad fishing trip, one that had yielded not a nibble.

"It's just—it's just depressing," he said in a voice as flat as the glassy harbor. "Man it is."

To grasp the depth of Hall's disappointment, you have to understand where he was coming from: fall 1989, the last time he went salmon fishing along the shoreline in Harrisville. Like so many of eastern Michigan's coastal towns back then, Harrisville swarmed each fall with shoreline anglers plucking chinook out of the lake with an almost metronomic regularity, as if the fish were coming off a General Motors assembly line. Reeling in salmon back then was indeed blue-collar sport—all you needed was a pole and a patch of public shoreline.

Hall remembered the carnival of the harbor parking lot. Beer flowed; car radios blared. But what he recalled most vividly was the crisp fall air tinged with the scent of burning hardwood from a nearby salmon smokehouse. And the old guy who filleted mounds of bronze carcasses for a dollar apiece—slicing and slinging fish skins into garbage cans with what seemed to be one fluid motion. Hall and a friend rolled home to Flint that 1989 day, arms sore from fighting all the 20-pounders, the back seat of their Geo Metro folded down to make space for coolers loaded with pale orange fillets. Hall moved away soon thereafter and had only recently returned to Michigan to care for his aging mother. He figured one upside to landing back in his home state was that he could once again hit Lake Huron's fall salmon run.

But now that the day had finally come, he didn't feel any of the old exhilaration. All he felt was foolish. On the drive up from Flint, Hall had worried aloud to his brother-in-law that the shoreline might be too packed to find a good spot. When they arrived, the sprawling boat launch parking lots were empty. Not a single angler could be found on the shore. Only a lonely pontoon boat puttered about the harbor. The fish fillet station was gone. So was the old bait shop. The only smoke in the air was from Hall's sad cigarette exhalations.

"This used to be the hot spot," he kept trying to convince his brother-in-law. "It used to be. It really did!"

After less than two hours they grabbed their gear and headed for the parking lot feeling as hollow as the three coolers in the back of the Dodge minivan. The lake of Hall's memory is dead, its salmon all but vanished in the past decade. The crash happened so fast that fisheries biologists have likened it to driving off a cliff.

Lake Huron was not hit as hard as Lake Michigan in the salmon crash of the late 1980s and early 1990s. But a decade later, it began to suffer its own devastating chinook demise. And this one looks to be permanent. It has become clearer with each passing year that the salmon program that created an economic boom in Lake Huron's coastal towns almost a half century ago was but a biological flicker—the brief several decades when humans pulled the levers on the ecological balance of a Great Lake and got the precise results they desired.

A big reason for the crash is that the salmon began breeding in the wild at unsustainable numbers—there were simply too many chinook mouths and not enough alewife tails. Tanner and his colleagues always knew Pacific salmon might figure out how to reproduce in Great Lakes tributaries, most likely in the cold, clear streams of Canada. And in the first decades of the stocking program it was known that some salmon had done just that. But nobody knew to what extent.

Biologists can tell which fish are stocked and which are "wild" because they clip a fin on each hatchery-raised fish, or mark it in some other manner. They can then estimate how many wild fish are swimming in a lake by comparing the number of fish caught from the hatchery population to the number of wild fish caught. By the time they started doing this ciphering in 2000, they were stunned. They had figured natural reproduction accounted for maybe 15 percent of

Lake Huron's chinook population but learned that, in addition to the 3.5 million hatchery chinook planted annually, nature was churning out as many as 16 million more. In other words, the biologists had their estimates almost completely backward—only about 20 percent of the lake's chinook population was hatchery raised.

Like pilots suddenly realizing their aircraft is loaded with more cargo than physics will allow it to stay aloft, fishery managers slashed their chinook plantings. It was too late. The finely calibrated salmon machine the biologists thought they had built had gone . . . wild. Alewife numbers on Lake Huron started to plummet around 2003, but few noticed it at first. The salmon fishing had been as good as ever—record catch rates were recorded just the year before. But it turned out this was not the sign of a healthy salmon fishery. It was a sign the salmon were running out of alewives to eat and starving in a manner that made a fisherman's lure irresistible.

Compounding the problem was that the alewives themselves began to run out of food due to an unexpected plummet in plankton populations tied to a surge in exotic mussels on the lake bottom. So the alewives were getting it on both ends of the food chain—gobbled up from above by too many salmon and starving from below because of a vanishing food supply.

Biologists doing netting surveys found empty chinook stomachs only occasionally in 2003, but in 2004 it was common. By 2005, the chinook catch from key ports on Lake Huron had plummeted from over 104,000 just three years earlier to 11,700. By 2010, the harvest crashed to barely 3,000, and it has since shown little sign of rebounding.

Few biologists today fault Tanner and Tody for so brazenly reconstituting the Great Lakes; Michigan's new research vessel for Lake Huron bears Tanner's name. "They had nothing to lose. The lakes were so destroyed that there was no place to go but up," said Michigan Department of Natural Resources biologist Dave Fielder. "It was a creative, clever, smart shot from the hip, because they didn't know what was going to happen. But boy, did it sure work out"—in terms of controlling the alewives, and kindling public enthusiasm for the Great Lakes. Fielder said it is hard for people today to grasp the significance of what Tanner and Tody did, and it wasn't just about fish.

"Back in the 1960s, nobody really cared much about the Great Lakes. They started to care when the salmon came on," said Fielder, who was a child in Michigan at the time of the first salmon introductions. He was aware of the program because he said those fish were pretty much the only good news on television and in the newspapers in the late 1960s.

"That was the first time I saw social optimism and excitement, and it was because of a fish," Fielder said. "That taught me that there are times that, as a society, we can be hopeful and excited."

Now the lesson Fielder and many other Great Lakes biologists have taken from Huron's recent collapse is one of humility, on the far end of the management spectrum where Tanner and the salmon pioneers tinkered.

"As humans, we always want to be in control, but we can't control the Great Lakes," said John Dettmers, a biologist with the Great Lakes Fishery Commission. "We changed it, but we didn't necessarily control it."

IN WISCONSIN, IN THE VERY MONTH THAT MICHIGAN'S JAY HALL was despairing what had become of the salmon fishery on Lake Huron, a chinook salmon factory was thrumming near the Lake Michigan shoreline. *Thump. Whoosh. Phst. Squirt. Stir. Bam.*

Although Lakes Michigan and Huron are actually one giant body of water they are in many ways distinct. Lake Michigan is a more "productive" lake due to the nutrients flowing from its tributaries that yield more alewife-sustaining plankton. The lakes also differ in water chemistry, temperature, depth and spawning habitat. And although Lake Michigan also is now home to a naturally reproducing chinook population, these fish are not breeding to the same degree that they were on Lake Huron in the years leading up to the salmon crash.

But even by 2014 signs of a Lake Michigan alewife collapse were beginning to appear; annual fishery numbers showed their overall numbers at an all-time low and older and larger alewives were disappearing. This was precisely what happened on Lake Huron just before the crash. Lake Michigan fishery managers nevertheless figured the best way forward was to keep planting salmon.

At the head of Strawberry Creek near Sturgeon Bay on the Door Peninsula, the salmon production line started in a manmade pool swarming with hatchery-raised chinook. These adult fish had spent their youth at a hatchery in fall 2011 and were then planted into a concrete pond at Strawberry Creek in spring 2012 for several weeks so the scent of "home waters" could be imprinted on their olfactory system. Then the gates to the pool were opened and the pinky-sized fingerlings flitted for the open waters of Lake Michigan.

Two and a half years later, these now log-sized beasts have followed their noses toward the Lake Michigan coast, into the manmade Sturgeon Bay Canal, up Strawberry Creek, and right back into the giant concrete pond of their youth. Here, on the last day of their lives, and the first day of their offspring's, a crane scooped them from the pool in writhing clusters. They were dumped into a tub bubbling with a numbing gas and then, one by one, fed down a chute.

Thump—The hammer on the SI 5 M3 Stun Machine delivered a knockout blow to the skull and out the back of the contraption came a limp Chinook, eye cocked toward the clouds with all the life of a button.

Whoosh—Rubber-gloved fishery workers slid the carcass onto a scale where it was weighed, measured for length and identified as male or female.

Phst—Females had their stomachs popped with the needle from a hose hooked to a carbon dioxide tank and were gassed up until their bellies excreted a stream of bright orange eggs into a bucket—about 5,000 per fish.

Squirt—Males were bent and squeezed by an assembly line worker in a manner that shot their milt into little plastic cups with dart-throwing precision.

Stir—A gloveless technician dumped a cup of milt into a three-gallon bucket containing the eggs from two females, and then dipped his fingers, sterilized by iodine, into the icy concoction and swirled until it foamed.

Bam—10,000 eggs were fertilized.

The crop of soon-to-be-fish—more than 200,000 on this day alone—was then taken by van to a state hatchery where the eggs would hatch in about six weeks. The little fish would feast briefly on nutrients in their egg sacs and be gobbling hatchery pellets by January.

By spring they will have grown to nearly four inches and be brought back to the concrete pool at Strawberry Creek for a few weeks of acclimation before the gates open and they wiggle along with the current toward Lake Michigan, where they will chase the lake's diminishing alewife population until they return to their concrete box two and a half years later. Or that's the plan.

"It's kind of scary, if you look at what happened on Lake Huron," Nick Legler, Great Lakes fishery biologist for the Wisconsin Department of Natural Resources, said during a break on the final day of the 2014 egg harvest at Strawberry Creek. "That's why we reduced stocking, to try to get the system back in balance."

States bordering Lake Michigan in 2016 stocked only about 1.8 million chinook—about half of what was planted just a few years earlier and well below the peak of about 8 million in the late 1980s.

Tanner wondered if any of it will do any good, because the alewife numbers continued to drop despite the slashes in salmon planting. "I'm worried about Lake Michigan," he says. "It's on the same course as Lake Huron."

Tanner's singular vision a half century ago almost instantly restored an ecological balance to the lakes and turned their forlorn waters into one of the world's most popular recreational fishing destinations. But patching up the top of the food web by stitching into it an exotic predator proved to be a relatively simple fix.

Now, it turns out, the lakes are suffering a second wave of invasions that make solving the lamprey and alewife infestations seem easy. This time there is no obvious remedy, because this time the food chain is being attacked at its most vulnerable point: at the bottom.

Chapter 4

NOXIOUS CARGO

THE INVASION OF ZEBRA AND QUAGGA MUSSELS

The first day of June 1988 was sunny, hot and mostly calm—perfect weather for the three young researchers from the University of Windsor who were hunting for critters crawling across the bottom of Lake St. Clair. Sonya Santavy was a freshly graduated biologist aboard a 16-foot-long runabout as the whining outboard pushed the boat toward the middle of the lake that straddles the U.S. and Canadian border.

On a map, Lake St. Clair looks like a 24-mile-wide aneurysm in the river system east of Detroit that connects Lake Huron to Lake Erie, and that is essentially what it is. Water pools in it and then churns through as the outflows from Lakes Superior, Michigan and Huron tumble down into Erie, then continue flowing east over Niagara Falls into Lake Ontario, and finally down the St. Lawrence Seaway and out to the Atlantic Ocean. The current pulsing through Lake St. Clair is so strong that if you were to hop in an inflatable raft at the top of the lake you'd flush out the other side in about two days—without having to paddle a stroke.

Water rushes so quickly through Lake St. Clair because it is as shallow as a swimming pool in most places, except for an approximately 30-foot-deep navigation channel down its middle. The U.S. Army Corps of Engineers carved that pathway in the early 1960s as part of the Seaway project to allow oceangoing freighters to sail between Lake Erie and the lakes upstream from it. When water levels were low or sediment high, sometimes that channel still wasn't deep enough, forcing ships to lighten their loads to squeeze through. This often meant dumping water from the ship-steadying ballast tanks—water taken onboard outside the Great Lakes. Water that could be swarming with exotic life picked up at ports across the planet.

As Santavy and her University of Windsor colleagues puttered over a rocky-bottomed portion of Lake St. Clair in the early summer of 1988, she whimsically dropped her sampling scoop into the cobble below. She was hunting for muck-loving worms but figured she'd take a poke into the rocks below because—well, to this day, she still doesn't know. "I can't even explain why it popped into my head," Santavy told me. "I thought—if we get nothing, we get nothing, and I'll just mark it off that this is not an area to sample."

Up came a wormless scoop of stones, the smallest of which were not much bigger than her fingertips. But there was something odd about two of those tinier pebbles. They were stuck together. She tried to pull them apart but she couldn't. Then she realized that one of them wasn't a pebble at all. It was alive.

NOBODY GAVE IT MUCH THOUGHT AT THE TIME, BUT IN THE YEARS following the Seaway's opening in 1959, species not native to the Great Lakes, ranging from algae to mollusks to fish, started turning up at a rate never before seen. In the Seaway's inaugural season it was the humpbacked peaclam native to Europe and Asia. In 1962 came *Thalassiosira weissflogii*, a single-celled alga capable of both sexual and asexual reproduction and, unlike sea lampreys, incapable of being controlled ecosystem-wide by any human measures.

Five more exotic species of algae showed up during the next two years and a tubificid (lake bottom–burrowing) worm native to the Black and Caspian Sea basins arrived in 1965. A water flea from Europe turned up the year after and a European flatworm two years after that. A crustacean native to the Black and Caspian Seas arrived in 1972. Three more exotic species of algae turned up the following year. And the alien organisms continued to arrive, year after year, with an almost metronomic predictability—all the way up to that steamy Wednesday morning on Lake St. Clair in 1988.

Santavy showed a fellow scientist aboard the research boat her living "stone," its wavy bands having allowed it to blend into the rocks she found it lurking among. It was obvious to both of them that it was some kind of clam or mussel, but the dime-sized mollusk looked like nothing Santavy's colleague had ever seen. This was odd. He was a graduate student whose job was to study freshwater clams of North America. This made Santavy suspicious enough to bring her specimen back to campus.

"I was just recently graduated, I wasn't really experienced and I didn't know a whole lot," she said. "But I thought: You know, maybe this is something . . ."

When Santavy returned to campus she showed her specimen to the professors in the lab. They were also flummoxed. They sent it to the University of Guelph outside Toronto, where an international mussel expert identified it as *Dreissena polymorpha*, the zebra mussel. This was not good news. The species, native to the Caspian and Black Sea basins, was well known on the other side of the Atlantic for its ability to fuse to any hard surface, growing in wickedly sharp clusters that can bloody boaters' hands and swimmers' feet, plug pipes, foul boat bottoms and suck the plankton—the life—out of the waters they invade.

The zebra mussel had already colonized rivers and lakes across Western Europe thanks to an extensive network of canals and locks that, like North America's Seaway, had allowed biological trouble to course through a continent like cancer cells in a bloodstream.

Hungary succumbed to an infestation in 1794, London in 1824. Rotterdam fell in 1827, followed by Hamburg in 1830 and Copenhagen in 1840. The mussels had spread to Switzerland, Finland and Italy by the 1970s. And then Santavy's specimen turned up in Lake St. Clair, some 3,000 miles from the closest known colony.

A zebra mussel has something of a "foot" that enables it to drag itself across a lake bottom, but even the fastest adult zebra mussel can only trundle along at maybe 14 inches per hour. A pioneering colony of the mussels also could not have inched its way, generation by generation, across the ocean and up the Seaway because the mussels would not have survived the ocean's salinity or depth. Scientists knew the most plausible way Santavy's mussel could have made the trip across the Atlantic and into the Great Lakes was in the friendly confines of a freighter ballast tank filled with water from a freshwater or semi-saltwater port.

Ship-steadying ballast used to be solid materials. In the 1800s bars of iron were used to balance schooners in the slave trade and Europe-bound ships laden with tobacco returned to the New World with bricks for ballast. But as freighters shed their sails and wooden hulls, acquired steam engines and grew to titanic proportions, the ships demanded ever-more stabilizing weight, particularly when a vessel was sailing with less-than-full cargo holds, unevenly loaded freight or through violent seas.

Naval architects soon realized water, at just over eight pounds per gallon, is plenty heavy to function as ballast. More importantly, it does not have to be manually loaded. It can be pumped in or out of the network of tanks tucked under the steel skin of a modern freighter.

But liquid ballast does have one huge drawback—it is anything but dead weight.

The discovery of Santavy's single shell might have meant little initially to the young researchers who found it. But seasoned ecologists knew the doom it foretold, like radiologists spotting a telltale speck on an X-ray; the important thing about the zebra mussel is to not consider each one as an individual organism but instead, like a cancer cell, part of a greater scourge that metastasizes as fast as currents flow. And, unlike some places in Europe and in the mussels' native range, North America has no natural mussel predators to keep their numbers from exploding in a manner never before seen.

Each female can produce one million eggs per year. Those microscopic offspring—called veligers and as small as a 10th of a millimeter in diameter—are covered with little hairs that help them catch currents and waves and "swim" to new locations during the first few weeks of their lives. The hairs also allow a baby mussel to snag food and begin to grow a shell, which eventually weighs it down and forces the mussel to settle on a lake or river bottom. There, it begins its blind hunt for a hard surface—rocks, glass, pilings, even other mussels—to attach to. Within a year those babies are sending out puffs of their own veligers to establish new colonies.

Distressing as the news was that the zebra mussel had made the jump across the Atlantic, nobody should have been surprised. As early as the late 1800s, naturalists had recognized the zebra mussel as an invasive species juggernaut. "The *Dreissena* is perhaps better fitted for dissemination by man and subsequent establishment than any other fresh-water shell," English zoologist Harry Wallace Kew wrote in 1893. "Tenacity of life, unusually rapid propagation, the faculty of becoming attached . . . to extraneous substances, and the power of adapting itself to strange and altogether artificial surroundings have combined to make it one of the most successful molluscan colonists in the world."

Another warning came in 1921 from Charles Johnson, curator of the Boston Society of Natural History. "The possibility of . . . the zebra mussel being introduced [to the United States] is very great. There is entirely too much reckless dumping of aquaria into our ponds and streams. A number of foreign freshwater shells, etc., have been introduced this way. Why not the mussel?"

Yet another alarm came in 1964, five years after the opening of the Seaway. "There is the real possibility that *Dreissena* polymorpha will eventually become established in the North American continent," cautioned biologist Ralph Sinclair.

A final warning came in 1981 when a group of scientists took the time to see what might be lurking in the ballast tanks of overseas Seaway freighters bound for the Great Lakes. They found the tanks were basically floating ecosystems, swarming with life sucked up from ports across the globe. The researchers specifically mentioned zebra mussels among the primary threats to make their way into the lakes by hitching a ride in ballast water, which is often discharged when an overseas ship arrives in the lakes in exchange for cargo. The U.S. and Canadian governments did nothing with the information.

The next year, in 1982, overseas ships were blamed for bringing in the spiny water flea that has since ravaged Great Lakes' native zooplankton and helped deplete stocks of little native fish that depend upon them. The year after that, Seaway ships were identified as the culprit for the arrival of another exotic tubificid worm. In 1986, an invasive fish from the other side of the Atlantic called the Eurasian ruffe was discovered in Lake Superior. Two years after that, in 1988, news of Santavy's zebra mussel find reached the public with a front-page story in the *Windsor Star*. It declared that a novel "zebra clam" might cost the region millions of dollars because of its ability to cling to hard surfaces and clog industrial water intake pipes.

"This guy hitchhiked inside a ballast tank," Paul Hebert, director of the University of Windsor laboratory where Santavy worked, told the reporter. "It's crazy to go on studying and studying this—we have to do something. We're getting new species in the lake all the time." The problem was regulators' hands were tied—by the Clean Water Act itself.

EDITORS AT THE CLEVELAND *PLAIN DEALER* HAD FINALLY HAD enough in the summer of '68 when the feculent Cuyahoga River that feeds Lake Erie did the unthinkable. "We have shown conclusively that the filthy condition of our river is a disgrace to the city," the newspaper's editors wrote, explaining how the surface of the river had in recent years become a viscous mixture of industrial chemicals and oils primed to ignite, threatening the warehouses, mills, machine shops, grain elevators and lumber yards lining the river banks. Of course, then it happened. "All of our warnings have had no effect and this morning, as a tug was passing down the river," the newspaper fumed, "a single spark from her smokestack fell upon the waste oil and literally the Cuyahoga River was set on fire!"

You may think you know this story. You probably don't. This was not *the* Cuyahoga River fire of the late 1960s, the one that burned in newspapers across the country and prodded Congress into passing sweeping federal pollution rules to heal the nation's waters that had been ravaged by a surge in post-war industrial pollution.

This fire did indeed spark just after the war—the Civil War. Like the famous blaze of 1969, this 1868 fire did little property damage and took no human lives. And like the flames of 1969, the 19th-century fire led to demands by the newspaper that local leaders "rid the river of this nuisance at once."

The government declined to act. Nor did it act when a flood through the city's industrial core unleashed flammable wastes that set the Cuyahoga ablaze again in 1883. Not much changed after a fire in 1912 either, even though five boat mechanics were immolated when a gasoline slick erupted into a "shriveling blast of blue flame." Cleveland's river burned again through the 1920s, 1930s and 1940s, yet never hot enough for politicians to curb the absurdity by holding the polluters accountable. A 1952 Cuyahoga River inferno did prompt a front-page headline in the *Plain Dealer*, though it focused on the fact a shipyard had burned down, not that a river had set it ablaze.

Cleveland was not the only Great Lakes city to make an industrial mess of a river. Similar fires blazed on tributaries from Chicago to Detroit to Buffalo, beginning in the late 19th century and stretching deep into the 20th century. The chemical dumping that caused these fires occurred with impunity because the federal water pollution laws that existed up into the 1960s were toothless to stop industries and cities from treating rivers as liquid landfills; civil and criminal penalties for polluters were basically nonexistent.

It was into this regulatory abyss that a spark dropped from a train passing over the Cuyahoga River on the morning of June 22, 1969. The flames were put out in a matter of minutes and merited only a small news item on page C-11 of the *Plain Dealer*. But then, like drifting embers, word of the fire landed in newspapers across the country, and a national outrage flared.

Somewhere between the raging blaze of 1952 and this little fire of 1969, the notion that a river could burn in the richest country on the planet became ludicrous. Criticism came from media outlets as large as *Time* magazine and as small as the Janesville (Wisconsin) *Daily Gazette*, which seethed:

> The Cuyahoga River was beautiful once. It had clear, sparkling water once, and children could play on its banks. It had fish once, and greenery to shelter wildlife. Now it is a stinking, fetid cesspool—an open sewer running through the heart of one of the

great cities of the United States of America. And Cleveland is by no means alone. The same thing has happened all over this land. The pollution of our rivers and lakes is more than just shocking. It is more than a disgrace. It is a damned outrage and a crime against the people of the United States. The rivers do not belong to the industrialists. They belong to the people—all of the people. When a river or a lake is polluted, it has been stolen from the people just as surely as if it were done at gunpoint. It has been stolen from you, and your children, and your children's children.

As the heat turned up on Cleveland to clean up its mess, Cleveland's mayor blamed the state for not being tougher on the industries it was charged with policing. Ohio law at the time gave state regulators supremacy over water pollution enforcement, and since the factories and mills along the river had received discharge permits from the state, no matter how weak they were, the city maintained it was powerless to force those businesses to change their polluting ways. The state, meanwhile, threatened "drastic enforcement measures" against the city of Cleveland, absurdly claiming the fire was fueled by oily runoff from city streets and leaky sewers.

Federal environmental regulators who descended on the river in the days after the fire concluded the source of the problem was obvious to anyone with a set of eyes—or nostrils. "In an area of the river dominated by Republic Steel, Jones and Laughlin and U.S. Steel, the river abruptly becomes a mixture of oil, debris and a multitude of different chemicals and solids," observed the head of the U.S. Department of Interior's Lake Erie office. He noted the color of the river upstream of those companies' belching pipes was a murky green. Downstream it was a chocolate froth that reeked of sulfur and ammonia.

The U.S. Secretary of the Interior called for public hearings that September, which only further highlighted the fecklessness of existing pollution rules. Republic Steel officials on that day refused to answer any questions. A U.S. Steel spokesman said he simply did not see the problem as his. "To the best of our knowledge," he said, "we are in complete conformity with all existing federal and state codes relating to water quality in the Cuyahoga River."

Nobody was fined a dime.

In 1972, Congress overrode a President Nixon veto and approved a sweeping package of amendments to the existing federal water pollution regulations that are known today as the Clean Water Act. This turned the tables by establishing the principle that industry does not have a "right" to pollute and must therefore apply for a permit to do so.

To get a permit, a company had to agree to install the best available waste treatment systems for the pollution it discharged. These permits had to be renewed every five years, the idea being that the volume of pollution a business could discharge would be continually ratcheted down as better treatment technologies inevitably evolved over the years and decades. Permit violations carried fines that could total tens of thousands of dollars per day and, just as the editorial writer in Janesville, Wisconsin, had pined for in the days after the 1969 fire, major offenders could be sentenced to jail.

The goals of the Clean Water Act were impossibly high—zero pollution discharges by 1985, with an interim target to make all the waters of the United States swimmable and fishable by 1983. The Clean Water Act missed those marks, but the improvements it brought have been immense. In the early 1970s two-thirds of America's lakes, rivers and coastal waters were unsafe for fishing or swimming. By 2014 that number had been slashed in half. Today there are some 60 species of fish swimming in the Cuyahoga, including water quality–sensitive sport fish like steelhead and northern pike.

But the Environmental Protection Agency left one huge loophole in the law the year after it was passed when it expanded an exemption for water discharges from military vessels to all ships sailing in U.S. waters. The agency was likely motivated by the notion that without a ship discharge exemption, its regulators could be on the hook to somehow police millions of recreational boats. Whatever the reason, the agency clearly did not see freighter discharges as a threat.

"This type of discharge generally causes little pollution," the EPA explained when it published the regulation creating the exemption, "and the exclusion of vessel wastes from the permit requirements will reduce administrative costs drastically."

But it would cost the Great Lakes dearly. As the zebra mussel infestation of the Great Lakes would make it abundantly clear, biologically contaminated ballast water is the worst kind of pollution because it cannot be fixed by plugging a pipe or capping a smokestack. It does not decay and it does not disperse. It breeds.

SANTAVY HAD FOUND ONLY ONE MUSSEL THAT SUMMER DAY IN 1988. Everybody knew there had to be more. They just didn't know how many. Tom Nalepa, then an ecologist with the National Oceanic and Atmospheric Administration, remembers making the three-hour drive from his office in Ann Arbor, Michigan, to London, Ontario, in March 1989 to meet with 11 other scientists about this latest Great Lakes invader. It turned out to be what is today known as the first International Conference on Aquatic Invasive Species, which has become an almost annual event that draws hundreds of researchers from across the globe. But the meeting on this chilly day wasn't called anything so grandiose. It wasn't called anything at all. It was just a dozen smart but mystified U.S. and Canadian scientists trying to share everything they knew about an organism that was spreading faster than their ability to read up about it.

The researchers in the room that day, in fact, couldn't even agree whether to call it a clam or a mussel. Conference host Ron Griffiths of the Ontario Ministry of Natural Resources took the Canadian-nice tack of referring to it as the "zebra mussel clam." The problem was there was almost no North American literature on the lifecycle of the zebra mussel because, until then, there had been no North American zebra mussels. The scientists had been gleaning what they could from research papers written in Russian, Polish and Danish just to figure out things like its preferred habitat, its temperature tolerance and its reproduction rate.

"A lot of the literature I've read is in another language, and I can only go as far as the abstract," Gerry Mackie, a mussel expert from the University of Guelph, confessed at the outset of the conference, a grainy video tape of which has survived more than a quarter century. The researchers turned on a carousel slide projector to look at how far the zebra mussels had spread since Santavy dropped her scoop to the bottom of Lake St. Clair just 10 months earlier. The room got quiet as the wheel stopped on each new image.

- An engine block found on the bottom of Lake St. Clair so encrusted with zebra mussels its piston holes were plugged.
- A Coast Guard buoy hauled in from Lake Erie so coated with shells it was unrecognizable.
- A Great Lakes beach littered with bleached mussel shells lying open on their sides, like so many little mouths.

Then Griffiths turned on a videotape of a mussel-smothered ferry wharf on the Canadian side of Lake Erie. There were so many shells nobody tried to calculate how densely they coated the wharf's pilings. It would have been like counting stars from the deck of an ocean freighter on a moonless night. "Man," ecologist Nalepa remembers thinking as he sat with his colleagues around tables littered with coffee cups and jars of zebra mussel specimens. "Nothing is going to be the same. Nothing."

There was some talk that day about how the plankton-gobbling mussels might affect native fisheries higher up the food web. But the scientists mostly worried about what the mollusks could do to the region's industries, given their ability to gum up pipes. Unlike mussels native to the Great Lakes, a zebra mussel can fix itself to hard surfaces by excreting plaque from a gland in the foot that it uses to drag itself along a lake bottom. After depositing this drip of glue, the gland then excretes proteins that trickle down the foot, fuse with the plaque and harden into remarkably tough, leathery threads. One adult mussel can spin more than 500 of these "byssal" tethers, creating a cement-like bond as durable as any epoxy you can find on a hardware store shelf.

Researchers quickly realized that water intake pipes used by cities and industries would likely be prime zebra mussel habitat; the hard surface inside a pipe provides an ideal place to attach and the constant flow of water—and the plankton floating in it—make for an easy meal, like a floating buffet. It was already starting to happen.

A biologist with a Michigan power utility popped in another video tape that showed that the zebra mussels were beginning to gather in plum-sized clusters on the water intake pipes at the Monroe Power Plant on the western shore of Lake Erie. He predicted that, unchecked, the mussels could cost the plant hundreds of thousands of dollars in chemicals and staff hours to poison and remove—*per day*.

Griffiths, the conference host, told the group that one of his biggest challenges was conveying to the public the severity of the situation. "Right now," he said, "there is no way anyone believes that a critter that small can cause any kind of problem." By the following December, some 50,000 people in and around the city of Monroe, Michigan, which draws its drinking water from Lake Erie, lost their water supply for more than two days when mussels and ice plugged an intake pipe nearly three feet in diameter.

The North American zebra mussel problem was made worse by the fact that they have no worthy predators in the Great Lakes, and in the most heavily infested areas, they soon began to cluster atop each other like gnarled coral at densities exceeding 100,000 per square meter. Each adult mussel, which typically grows no bigger than a nickel, can filter up to a liter of water per day, sequestering inside its hard little shell all the nutrients contained within that water.

Razor-sharp mussel shells littering a Great Lakes beach.

By the end of 1989, zebra mussels had turned up all across the Great Lakes, west to Duluth, south to Chicago, and east to the St. Lawrence River below Lake Ontario. A colony was also found near the head of the Chicago Sanitary and Ship Canal that provides a manmade connection between the Great Lakes and the Mississippi River basin. That meant the mussels now had access to a watershed that spans almost half of the continental United States.

But the most ominous mussel development of 1989 made no headlines. Researchers on Lake Erie found what appeared at first to be a slightly different version of the zebra mussel. It was, they would learn two years later, the quagga mussel, named after a subspecies of actual zebras that went extinct in the 1800s. All that remains of the African savanna grazers are seven skeletons, including one on display at University College London. But their molluscan namesake today, in the Great Lakes alone, numbers in the quadrillions.

Zebra mussels proved to be an expensive nuisance indeed for industries and cities that depend on water, costing billions of dollars over the past quarter century to invent, build and maintain treatment systems that use things like chemicals, heat and UV light to keep pipes open and water flowing through everything from nuclear power plants to kitchen faucets. Yet the ecological damage wrought by zebra mussels is minor in comparison to their cousin, the quagga mussel. Unlike zebra mussels, which typically aren't found at depths beyond 60 feet, quaggas have been plucked from waters as deep as 540 feet. This depth tolerance, coupled with the fact that quaggas don't require a hard surface to attach to, means they can blanket vast swaths of lake bottom inaccessible to zebra mussels. Zebras also only feed during the warmer months. Quaggas filter nutrients out of the water year-round.

In 1992, three years after quagga mussels were discovered in Lake Michigan, zebra mussels still made up more than 98 percent of the lake's invasive mussel population. By 2005 that relationship had completely flipped, with the quaggas making up 97.7 percent of the invasive mussel population and smothering the deepwater lakebed in a manner zebras never could. Although the waters of Lake Superior lack levels of the shell-building calcium that zebra and quagga mussels require to thrive, the mussel impacts on Lake Michigan have been similarly repeated on the other lakes, particularly Huron and Ontario. The chaos this has brought is like nothing—not even the sea lamprey—the lakes have suffered in their 10,000-year history.

THE PUBLIC CAN COMPREHEND THE DEVASTATION OF A CATA-strophic wildfire that torches vast stands of trees, leaves a scorched forest floor littered with wildlife carcasses and turns dancing streams into oozes of mud and ash. But forests grow back. The quagga mussel destruction is so profound it is hard to fathom.

"People look at the lake and don't think of it as having a geography. It's just a flat surface from above—and from there it looks pretty much the same as it did 30 years ago, but under water, everything has changed," University of Wisconsin–Milwaukee ecologist Harvey Bootsma says.

The mollusks now stretch across Lake Michigan almost from shore to shore. People might still think of Lake Michigan as an inland sea full of fish. It's more accurate to think of it as an exotic mussel bed sprawling across thousands of square miles. Lake Michigan's quagga mass in one recent year was estimated to be about seven times greater than the schools of prey fish that sustain the lake's salmon and trout. Under some conditions the plankton-feasting mussels can now "filter" all of Lake Michigan in less than two weeks, sucking up the life that is the base of the food web and making its waters some of the clearest freshwater in the world.

Just how much have things changed since quagga mussels took over? A simple way to gauge the amount of plankton in a water body is to take a visual sounding using a crude device called a Secchi disk, named after a 19th-century Italian priest tapped by the one-time Papal Navy to take water clarity readings in the Mediterranean.

The disk is, typically, an eight-inch diameter metal plate with four equally sized alternating black and white wedges, almost like a monochromatic version of the yellow and black nuclear fallout shelter sign. It is lowered by rope into a water body and the point at which it disappears is the water's Secchi depth. In the late 1980s, before the mussels blanketed the lake bottom, Lake Michigan's average Secchi depth was 6 meters, or about 20 feet. By 2010 the average depth had tripled and readings began coming in at beyond 100 feet. This nearly vodka-clear water is not the sign of a healthy lake; it's the sign of one in which the bottom of the food web is collapsing.

One study on southeastern Lake Michigan revealed that by 2009, phytoplankton levels in springtime—the prime plankton-growing time of year—had dropped nearly 90 percent since the mussels took over the lake bottom. It's probably not a coincidence that the lake's fish populations have dropped at the same time.

Annual trawling surveys show the lake's biomass, or overall weight of prey fish, has plummeted from an estimate of about 350 kilotons in the late 1980s to barely 5 kilotons by 2014. And then a federal fisheries survey crew went fishing one warm September day in 2015.

IT WAS THE FIRST DAY OF FALL AND THE SUN WAS STILL RISING on a Lake Michigan that was flat as a frozen pond when the fishing boat captain's command crackled over the loudspeaker as his crew prepared to drop a net off the back deck: "Spit on it."

The winch operator smiled and obliged, licking the tips of his fingers and swiping them on the black webbing before he hit the lever to lower the fish net to the lake bottom more than 400 feet below. He explained to me that the captain reckons that if spit is good luck for kids with worms twisting on fishhooks, then it is good luck for the five-person crew of the $6 million *Arcticus*. It brought the wrong kind of luck over the next seven hours as the massive trawler dragged a 40-foot-wide net back and forth across the lake bottom, lawnmower-style.

This crew from the U.S. Geological Survey was not fishing for flesh. They were fishing for clues. The group was near the end of an annual three-week "prey fish" survey of the bottom of Lake Michigan that has been conducted every year since 1973. The purpose of these autumnal expeditions for bite-sized fish—diminutive species like sculpins, chubs and alewives—is to check the lake's gas gauge. This is because the Great Lakes, and Lake Michigan in particular, have been managed primarily for recreational fishermen for the past half century. The more of these little fish that research crews find swimming in the lakes, the more predator fish—hatchery-raised salmon and trout—that can be planted to gobble them up. The whole operation is kind of like the aquatic equivalent of an oversized hunting preserve stocked with trophy sized elk and deer.

The *Arcticus* team's surveying is far from a precise means of weighing how many pounds of salmon and trout food is swimming in the lake, though it is an exercise in fishing precision. Year after year, the Lake Michigan researchers, now guided by satellites, hit the exact same seven spots of lake bottom. They start in the waters off Michigan's Upper Peninsula at the northern end of the lake and then sweep clockwise about 300 miles down its eastern side, across the lake's U-shaped southern tip near Chicago and then back north along the western shore. Each survey site includes several sweeps of lake bottom at depths varying from beyond 400 feet to less than 60 feet. After each sweep the net is hauled to the surface so the catch can be analyzed.

This is a lot of lake bottom scraped clean of its fish but it doesn't even cover a sliver of the overall lake; Lake Michigan has a surface area spanning more than 22,000 square miles and in places plunges into

ink-black depths beyond 900 feet deep. Still, from the data—the fish—
pulled from the lake on these expeditions, the biologists, assisted by
computer models that have been refined over decades, are confident
they are able to sketch a good estimate of the number of pounds of
prey fish available in the lake in any given year. It's not unlike how a
political pollster who samples a mere 1,000 likely voters in a nation
of 300 million can be confident in his ability to identify presidential
front-runners. But more than putting a number on the pounds of prey
fish in the lake, the surveys are particularly useful in providing an estimate of the relative abundance of prey species from one year to the
next. In this sense, the surveys show biologists which direction things
are headed. And in recent years, if Lake Michigan's prey fish population were plotted on a graph like the Dow Jones Industrial Average, the
trend would look something like the Panic of 1929.

Consider the alewife. When the surveys started in the early 1970s,
biologists estimated that Lake Michigan harbored around 100 kilotons
of the Atlantic invader that was at the time the lake's dominant prey
fish, and the one most favored by the salmon that are the Great Lakes'
most popular sport fish. By 2004 the number had dropped to 14 kilotons. By 2014 it had plummeted to 1.6 kilotons.

And by the time I boarded the *Arcticus* to go fishing in the waters
about 30 miles north of Milwaukee in fall 2015, nobody was talking
kilotons. The crew had already hit five of the seven survey areas on its
annual tour, and when I asked the lead researcher how things were
going he shook his head. More scientist than fisherman, he said he
didn't want to give me a figure or paint a picture that wasn't precise.
But he said things did not look good.

The first haul that day broke from a Lake Michigan surface that
was black as oil because the water was so glass-flat that the sun had
neither wave nor ripple off which to glimmer. The catch was dumped
into a tub and hauled inside the boat so each fish could be identified,

counted, measured and weighed. There wasn't much work to do. I was expecting at least hundreds of pounds of writhing fish, but the whole catch weighed less than four pounds. We tried again at a slightly shallower depth. And then again, and again and again, and the net kept coming up with close to nothing.

"Gosh," the expedition leader said after one sweep of the lake bottom netted a single fish the size of my pinky.

"That's just embarrassing," said another crew member.

The *Arcticus* headed toward shore at the end of the day with a total catch that could fit in my first grade daughter's school backpack. I asked the expedition leader if he thought we just had bad luck: "What can I say!" he said with a pained smile and a shrug. Then he shrugged harder, like a sitcom character, scrunching his shoulders and holding his hands out, palms up. So I asked him if he had seen enough to say Lake Michigan's prey fish are crashing, particularly the alewives that sustain the salmon fishery. Such a crash has already happened on Lake Huron.

"I wouldn't say we've crashed," he finally said, "but at least we're in the process of crashing, you might say."

NOT ALL FISH ARE STRUGGLING. THE INVASIVE ROUND GOBY, another Seaway interloper that arrived just a couple of years after the mussels and is also native to the Caspian and Black Sea region, evolved to feast on the flesh of quagga and zebra mussels by cracking their shells with molar-like teeth. Now this bug-eyed, thumb-sized fish is thriving across the Great Lakes, nowhere more so than on Lake Michigan, where the trawl surveys in recent years show it is among the most common forage fish. This means two species from the other side of the Atlantic now play lead roles in the drama unfolding under the waves. Call it the Caspianization of the Great Lakes. But this is more than a matter of invasive fish replacing native fish; the mussel impacts are now rippling ashore.

"People really don't grasp what has happened here," University of Wisconsin–Milwaukee ecologist Harvey Bootsma explained to me on a frigid early November day as he strapped on a scuba tank, climbed over the back of the boat and plunged to the lake bottom 30 feet below. He was only about 800 yards off the beach of a popular park in the leafy Milwaukee suburb of Shorewood. But he might as well have landed on another continent because under the surface Lake Michigan bears little resemblance to the freshwater wonder that left early European explorers awestruck with its teeming herring, trout, sturgeon, perch and whitefish. Down below, the lake has pretty much become just a goby show. They thrive on the invasive mussels amid a shin-high forest of a nuisance seaweed-like plant called *Cladophora*, which needs three things to thrive: sunlight, nutrients and a hard surface.

The mussels have provided all three. Their plankton-stripping ability has dramatically increased the depths to which sunlight can penetrate. Their shells provide a surface on which the seaweed can grow and the mussels' phosphorus-rich excrement fuels the plant's growth. The result is an endless forest of brilliantly green, hair-like tendrils swaying in the current, invisible to anyone on shore—until relatively small amounts of it break off, wash ashore and, along with the mussels it has attached to, rot.

The septic-smelling muck plagues some of the lakes' most spectacular shorelines, including Sleeping Bear Dunes, a 35-mile stretch of federally protected coastline on Lake Michigan's eastern shore. At Newport State Park on the other side of the lake, Wisconsin's only wilderness park, the sludge can get shin-deep. A park employee in recent years has taken to keeping a laminated picture of the beach from the pre-quagga days, to show visitors how pleasant her sandy shores used to be. But the mess on the beach is nothing compared to what's happening all across the lake bottom.

"What people see on a beach is just the tip of the iceberg," said Bootsma. "You've got maybe a few thousand square feet of it on the beach—but just offshore, out in the lake, you've got thousands upon thousands of acres of it."

Bootsma finds the changes professionally interesting, but personally distressing. He attributes his whole career to summer days he spent as a child on Georgian Bay in northern Lake Huron, fishing for native bass and perch and snorkeling to the rocky bottom to capture crayfish. He still remembers the pit he felt in his stomach when his family packed up from their camping trips to Killbear Provincial Park and drove three hours south to their home in the steel mill city of Hamilton, Ontario.

"We'd go up there for the holidays and I'd be out in the sun and on the water every day all day, and I remember every time when it was the day to go home, I'd be close to tears," he told me. "I still remember telling myself that when I grow up I'm going to get a job that will keep me on these lakes all the time."

Outside his office window at the University of Wisconsin–Milwaukee's School of Freshwater Sciences are the grain elevators and coal piles that define the city's inner harbor. That harbor is connected to Lake Michigan, which is connected to Lake Huron, which connects to Georgian Bay. They all have different names, but in actuality they are the same lake, and the largest lake by surface area on the globe. It is no longer the lake Bootsma fell in love with. He's known this in his head for years because of his almost weekly trips out to his research station on the bottom of Lake Michigan.

But the changes really didn't hit him personally until he traveled back to Georgian Bay's Killbear park a few years ago with his own children. "I snorkeled some of the same areas that I did as a kid. I still saw some bass and that made me happy," he said. "But it was just devastating to see all the mussels and the gobies."

What's even more distressing for him is the idea that his children don't even know what they're missing. Ecologists call it the "shifting baseline phenomenon"—a fancy way of saying that kids are getting cheated out of the lakes their moms and dads loved. "It's just so sad to see it changing so much," Bootsma said. "This isn't the lake it was 25 years ago, and it's probably not the same lake it's going to be in 10 years."

It's not just native fish species and summertime beachgoers that have been hit by this biological pollution. Invasive species can have effects just as toxic as the nastiest chemicals concocted in a lab. A textbook example is the botulism outbreaks that have been killing tens of thousands of birds on Lakes Michigan, Erie and Ontario in recent years. It's a crash course in living pollution that is as simple as it is frightening:

- Invasive mussels have increased water clarity.
- That has led to a bloom in the sunlight-loving *Cladophora* that eventually dies and burns up massive amounts of oxygen as it decomposes on the lake bottom.
- That has opened the door to botulism-causing bacteria that thrive in oxygen-starved environments.
- The invasive mussels, many biologists believe, suck up those bacteria and are, in turn, eaten by gobies.
- The poisoned gobies become paralyzed and are easy prey for birds like loons, grebes and gulls.
- The birds die.

This is not a rare occurrence. Biologists estimate more than 100,000 dead birds—including bald eagles, great blue herons, ducks, loons, terns and plovers—have piled up on Great Lakes beaches since the botulism outbreaks turned rampant in 1999.

EVEN AFTER CONTAMINATED BALLAST WATER WAS BLAMED FOR the zebra and quagga mussel invasion of the Great Lakes, the EPA continued to exempt ship ballast discharges from the Clean Water Act. Trying to fix the problem on its own, in 1990 Congress instructed the U.S. Coast Guard to begin regulating ballast water in overseas ships visiting the Great Lakes.

The Coast Guard, then an agency under the U.S. Department of Transportation, which also runs the Seaway, did not go hard on the freighter industry. Ships sailing from foreign ports were only *asked* to flush their ballast tanks in mid-ocean with saltwater to try to expel or destroy any freshwater invasive stowaways.

Yet the U.S. and Canadian governments could not simply order overseas ships to stop using ballast water while sailing in the Great Lakes without exposing the ships to extraordinary peril. Using ballast to balance a ship or stabilize it in rough weather is not just a matter of comfort for the crew and passengers, like shock absorbers on a car. It can be as critical to a ship's safety as properly functioning wing flaps are to an airliner.

Improper ballast management, in fact, played a role in the greatest catastrophe in the history of Great Lakes navigation, the sinking of the S.S. *Eastland*, the "Speed Queen of the Great Lakes," in 1915. On a drizzly July morning that year, some 5,000 employees of the Western Electric Company and their family members descended on the Chicago River near Clark Street in Chicago. They were there, decked out in formal attire, to board a fleet of vessels chartered to sail about 40 miles across the southern tip of Lake Michigan to a company picnic at the sand dunes near Michigan City, Indiana.

Passengers started boarding the *Eastland* at 6:30 a.m. and bounded up the gangway at such a pace that within several minutes the ship started to list to its starboard side, toward the wharf. This was not initially an unusual or alarming phenomenon. Nearly a football field in length but a mere 38 feet wide, the ship was built in 1903 to knife through the water at 19 mph, a blistering pace for the time. The velocity came at a price—the *Eastland* was notoriously tippy.

As she tilted toward the wharf, the *Eastland's* chief engineer ordered that the port-side ballast tanks be filled to balance the crush of passengers boarding on the starboard side, and by 6:51 the ship was steadied. Passengers continued to pour onto the ship at a rate of almost one per second. At 6:53 the ship began to list toward port. Ballast in the starboard tanks was ordered, and the slow motion wobbling stopped—briefly. At 7:10 the ship was filled to capacity—some 2,500 souls—and listing toward port again. A port side ballast tank had been ordered emptied but, tottering like a drunk, the ship continued to list. Then, around 7:20, as the gangplank was lifted, the tipping worsened. By 7:28 everything from dishes to a lemonade cart to a piano went crashing across the decks. With hundreds already below deck to escape the morning's foul weather, the ship rolled gently, almost whale-like, onto its port side and came to rest, half submerged, in the muck of the 20-foot-deep river.

Within a half hour those survivors who had jumped or been dumped into the river or scrambled onto the exposed starboard side had been brought safely ashore. Rescue attempts for those trapped inside the *Eastland* would drag on for hours. Working just feet from shore, rescuers used cutting torches to carve through the steel hull to try to reach any survivors who had found air pockets inside the ship. It turned out to be as much a recovery mission as it was a rescue mission.

"Fred Swigert, a city fireman, worked three hours lifting bodies from the hold. Then a diver passed up the body of a little girl, her flimsy dress a pitiful clinging shroud," the *New York Times* reported the next day. "Swigert placed the little body on a stretcher, and then, looking closely at the drawn features, gasped and fell unconscious across the body. It was his own daughter."

The sinking S.S. *Eastland* in 1915, which killed hundreds of passengers when it rolled onto its port side just off the docks in Chicago.

All tallied, 844 people drowned that day including, remarkably, only two crewmembers. It was the worst nautical disaster in the history of the Great Lakes and claimed more passengers than the *Titanic* sinking three years earlier, whose death toll included hundreds of crew members.

The two catastrophes are peculiarly related; the narrow-hulled *Eastland's* perilous hydrodynamics were made more so just weeks before she sank following a federal mandate that extra life rafts be lashed to her upper deck—a legislative legacy of the *Titanic* disaster. Even though the extra lifeboats might have made the top-heavy *Eastland* even more wobbly, when a grand jury indicted members of the crew later that summer, it specifically cited their "entire lack of understanding of the nature and proper use of water ballast" as a prime factor.

The same could be said nearly five decades later with the opening of the St. Lawrence Seaway.

IN 1993 THE U.S. COAST GUARD MADE EXCHANGING BALLAST water with mid-ocean saltwater mandatory, yet wave after wave of new invasions kept rolling into the Great Lakes. The problem was about 90 percent of the ships arriving in the Great Lakes from foreign ports at that time came fully loaded with cargo and therefore did not officially carry any ballast water, and the new law exempted those ships from the ballast exchange requirement. But just because a captain declared his vessel to be ballast-free did not mean his tanks were empty. Most tanks still carried loads of sludge—up to 100,000 pounds of it—along with thousands of gallons of residual ballast puddles that cannot be emptied with a ship's pumps.

Subsequent studies revealed these muddy puddles swarmed with millions of organisms representing dozens of exotic species that had yet to be found in the Great Lakes. Furthermore, the life lurking in this muck had an easy escape from the bowels of a ship once a captain unloaded cargo at his first Great Lakes port of call and then filled his ballast tanks before steaming to the next Great Lakes port. That ballast—and any organisms churned up from the sludgy bottoms during the cross-lakes voyage—could then be discharged when the captain swapped it for cargo at the next port.

Between 1990 and 2008, 27 new exotic species were discovered in the lakes. The pace peaked around 2005, when a foreign organism was being detected, on average, about once every eight months. The official tally is there are now at least 186 nonnative organisms swimming or lurking in the Great Lakes. Not every one of these life forms can be defined as "invasive" because some were planted and are considered by many to be an asset to the region (the salmon and some stocks of exotic trout, for example) and others apparently exist in their new home without any discernable negative environmental or economic impact. But sometimes it takes two foreign organisms working together to cause utterly unpredictable trouble, as in the apparent case with invasive mussels and gobies working together to trigger botulism outbreaks in native birds. These kinds of chain reactions can take years or even decades to unfold, and they make it impossible for biologists at any point to know which introductions will be harmless and which will become troublesome, if not disastrous.

The term "invasive species," in fact, is something of a fuzzy concept. It's basically an organism that has caused undesirable disruptions to the lakes' ecology. This includes fish like common carp (in the lakes since the 1880s), alewives and lampreys, as well as a host of smaller organisms brought in by Seaway ships, including zebra and quagga mussels, spiny water fleas, fishhook water fleas and the bloody red shrimp.

The disruptions that these ballast water invaders, many from the waters of the Black and Caspian Seas, have collectively wrought on the lakes' native species was enough for one of the Seaway's biggest boosters to finally, in summer 2000, blurt to a group of foreign shippers what biologists had been thinking for more than a decade.

"Your ballast water from the Black Sea is destroying our Great Lakes!" the late James Oberstar, a Minnesota member of the U.S. House of Representatives, recalled for me one steamy June day in his Capitol office in 2005. "It's that simple."

Three years and a couple of more ballast water invasions later (including a deadly fish virus), in 2008 the U.S. Seaway operators began requiring all Great Lakes-bound overseas vessels to flush even their "empty" ballast tanks with mid-ocean saltwater.

No new exotic organisms have been found in the Great Lakes since, a point shipping industry advocates tout. Although it is generally agreed that mid-ocean flushing is an excellent first step toward closing the door to new ballast water invasions, it is also generally agreed upon that ballast water disinfection systems similar to sewage treatment plants must be used by ships to provide ample protection for the Great Lakes and other U.S. waters. It's a numbers problem.

Even if a ballast flush kills or expels more than 99 percent of hitchhikers, ships arriving in the Great Lakes from ports around the globe are still far from sterile. One study has shown that a single freighter ballast tank can harbor some 300 million viable cysts of primitive dinoflagellates, which scientists dub the "cells from hell" because they can produce a deadly neurotoxin. So a flush that eliminates 99 percent of that ballast tank's inhabitants could still carry three million potential invaders. That's just one ballast compartment, and that's just one species.

Critics of the saltwater flushing-only practice also say it is naïve to assume the ballast problem has been solved because it's been years since a new invader has been detected in the Great Lakes. Sleeper colonies can lurk for years—even decades—before their numbers grow big enough for them to get noticed. There is also no organized survey under way to detect new invaders, so Great Lakes species discoveries also tend to be accidental—a fisherman hauls up a strange cluster of slimy critters coating his nets, federal researchers think they've stumbled on a school of baitfish only to look closer to realize they are a shrimp nobody had ever seen on this continent, or students doing a routine survey of a lake bottom stumble upon something their professor doesn't recognize—the precise way Sonya Santavy stumbled upon the first North American zebra mussel.

Jake Vander Zanden knows how tricky it can be to discover a new invasive species—not just in the Great Lakes but in relatively tiny inland lakes as well. The professor at the University of Wisconsin's Center for Limnology has an office within feet of the lapping waves of Lake Mendota. Limnology is the study of inland lakes, and the Center's work makes Mendota one of the most exhaustively studied water bodies on the globe. A day on Mendota for the UW's limnologists is about as predictable as it gets, like botanists exploring their own backyards. Usually.

On September 11, 2009, Vander Zanden took a boatload of undergraduates about a quarter mile out on the lake and dropped a net in the water to sample for plankton. The students pulled the net from the water and dumped its contents, as instructed, into a jar. Vander Zanden did a double take when he saw that jar was teeming with a type of aquatic flea that has a telltale barbed tail—one that makes them a notoriously difficult meal for Great Lakes native fish. It was the spiny water flea, yet another Seaway ballast invader.

The spiny water fleas likely arrived in Lake Mendota aboard a recreational boat towed over from Lake Michigan some 80 miles to the east. Nobody knows precisely when. But just months after their initial discovery, UW researchers started finding them at densities over 1,000 per cubic meter of water—the thickest concentration of spiny water fleas recorded anywhere in the world. Lake Mendota has 500 million cubic meters of water, so in just a matter of weeks researchers went from thinking that the lake had a population of zero spiny water fleas to calculating it could be home to as many as 500 billion.

The episode makes Vander Zanden dubious of the idea that just because no new species have been recorded in the Great Lakes since 2008, nothing new has invaded during that time. Or that all the species that arrived before 2008 have already been discovered.

"My concern," he said, "is that invasive species are able to fester at undetectable levels, then 'boom' when conditions become right."

And the bigger the lakes, the bigger the boom. Lake Mendota has a surface area of a mere 15 square miles. The Great Lakes span more than 94,000 square miles.

The EPA has a watch list of dozens of species that remain a risk to invade the lakes in ship ballast tanks, despite the saltwater flushing requirements. On that roster of the unwanted is the notorious *Dikerogammarus villosus*, otherwise known as the killer shrimp. The murderous crustacean got its name because it makes a mess of the ecosystems it invades by destroying its prey with vise-like jaws and then leaving its victims for dead, often without swallowing a bite. These shrimp, which can grow bigger than an inch, have been spreading through Europe's canal system for decades. Studies show they have a salt tolerance just slightly below what is typically found in the ocean, which means that if the shrimp get sucked into the ballast tank of a ship sailing for the Great Lakes, there is perhaps only a membrane-thin margin of protection against another wave of destruction in the lakes.

TAKING ADVANTAGE OF A PROVISION OF THE CLEAN WATER ACT that allows citizens to sue to require the EPA to enforce the law, conservationists took the ballast issue to court in 2001. It took more than a decade of legal tangling, but in 2011 the EPA finally agreed to mandate treatment systems for overseas ships discharging ballast in U.S. waters. The systems, which will use things like chlorine, ozone and UV light among other pesticides to kill ballast dwellers, will not be required of all ships until sometime after 2021.

But even with a decade-long phase-in period, the EPA decided to require treatment technology that will only begin to chip away at the problem. Under the rules, for organisms 50 microns or greater—the thickness of a newspaper page is about 70 microns—ships can discharge fewer than 10 living specimens per cubic meter of water. For smaller organisms—between 10 and 49 microns—the limit is 10 *million* per cubic meter. And each ship sailing into the Great Lakes can be carrying as much as 25,000 cubic meters of ballast water—roughly equivalent to 10 Olympic-sized swimming pools.

Although these treatment standards should reduce the amount of life that would normally be spilling into the lakes from ballast tanks, think of the problem like a campfire. The EPA's treatment requirements are a little like the first gallon of water you slosh on the fire at the end of the night. It might knock down the flames but it will take several more gallons to properly soak the embers to ensure you've snuffed their glow. Then it might take a few more gallons before you kill the telltale hiss and can go to sleep secure in knowing that your fire isn't about to become someone else's.

Rudi Strickler, a Swiss-born zooplankton expert with the University of Wisconsin–Milwaukee, says there are plenty of freshwater organisms that will indeed die when hit with a blast of saltwater, a dose of chemicals or a flash of UV light. But he considers these species pushovers.

Then he mentions tardigrades. The earth might be teeming with indomitable beasts, he explained, but nothing—not the polar bear, not the crocodile, not the anaconda—matches the ferocity of the microscopic tardigrade when it comes to doing whatever it takes to survive. Also known as "water bears" or "moss piglets," tardigrades can, and do, thrive across the globe, because they can survive almost anything the earth—or even an earthling—can throw at it. They can survive an oven broiler, and they can weather temperatures hundreds of degrees below zero. They can be found scrambling across mountainsides in the high Himalayas and mucking about in ocean canyons two miles deep.

Under a microscope, the pudgy eight-legged waddlers look so cuddly they could pass for a child's stuffed animal. They captured the imagination of the science world in 2007 when European researchers literally went to astronomical lengths to destroy them. They put a platoon of slumbering tardigrades on a rocket in Kazakhstan, blasted them into space and exposed them to everything the cosmos could throw at them. A person will wither within seconds if left unprotected in the vacuum of the universe; these tardi-nauts were left drifting across the hellaciously frigid, cosmically radiated heavens in an open satellite compartment for 10 days. "Then they brought them back," said Strickler, "and they woke up and walked around."

Not all the tardigrades survived, but among those that did were a group robust enough to have perfectly healthy offspring.

There are some 1,000 different species of tardigrades, some of which are so small they could be padding about on the period at the end of this sentence. Although none pose an obvious threat to the Great Lakes—some species of tardigrades already call the region home— Strickler said it's necessary to consider their tenacity when trying to draw up a ballast battle plan against nature's spectacular array of life.

"Don't pick the ones that you can beat up easily," he said. "Pick ones that know all the tricks of life."

The case of the tardigrade makes it obvious that it would be impossible to design a ballast treatment system that is 100 percent fail-proof. The goal is to get as close to that as possible, and critics say the new treatment standards simply don't.

The EPA itself commissioned a panel of scientists to assess whether treatment systems could meet standards that are 10, 100 or 1,000 times more stringent than those the EPA ultimately adopted, and the agency said the answer it got back was no. This may not be entirely true. Eight of the 21 scientists on the panel took the extraordinary step of writing the EPA to argue that the agency's conclusion based on their research was misconstrued, and that there are indeed systems available that could go far, far beyond those the agency required of the shipping industry, particularly if the EPA had considered forcing ships to unload their ballast to onshore treatment plants.

So after the new ballast regulations were announced in 2013, conservation groups sued yet again, claiming the EPA botched its obligations under the Clean Water Act by failing to require the best ballast treatment available. The litigants also argued that the EPA failed to establish a safe discharge level that would ensure protection of the waters, even if that level is so low that existing technology cannot achieve it at the time the permit is issued. The Clean Water Act is designed to set targets for industries to shoot for as treatment technologies evolve and permits are renewed every five years. And that is the conservationists' problem: the shipping industry was given no such target. Instead, a jargonish sentence in the EPA's 2013 ballast rules said only that ballast discharges "must be controlled as necessary to meet applicable water quality standards in the receiving water body . . ."

What this means is that ship captains are merely instructed to not launch any fresh invasions. It doesn't tell them what it will take to ensure that doesn't happen, even though it is impossible for a captain at the wheel of a 700-foot freighter to know what trouble he might be carrying into a port when the pollution is alive, capable of reproducing and in some cases smaller than a pinhead.

The conservationists' argument held sway with the United States Circuit Court of Appeals Second Circuit, which ruled in 2015 that the existing ballast discharge rules are inadequate. The court ordered the EPA to require tougher treatment standards for the shipping industry, though it gave the agency no deadline to impose those standards.

Solving this problem is far more complicated than plugging pipes spewing oil, or devising a poison that specifically targets pests like sea lampreys. In preparing the now-rejected ballast treatment standards, the EPA turned to some of the country's best scientists in the field to help establish a safe number of organisms that could be discharged per cubic meter of water while still protecting the Great Lakes and other U.S. waters from new invasions. The scientists said they couldn't do it.

The problem is there are many factors that go into an invasion equation beyond the number of organisms released, including the condition of the organisms when they are discharged, whether they are sexually mature and, if not, whether they are released into waters in which they might be able to survive long enough to become so.

The only thing the panel could agree on is that the fewer organisms allowed to survive in a ballast tank, the better. Beyond that they were at a loss because, they said, you can't just pick a magic number and call it safe.

Unless the number you pick is zero.

That is the number the public demanded in the early 1970s for the tolerable times a river should be allowed to ignite. And that is the number Isle Royale National Park Superintendent Phyllis Green aimed for when she learned in 2007 that an invasive virus deadly to dozens of freshwater fish species was creeping toward her rugged, forested island in the middle of Lake Superior. Green's focus instantly turned to the island's coaster brook trout—a beleaguered native species that once numbered in the millions in Lake Superior but is now counted by the hundreds. "If you have only 500 fish and you have a disease that can kill fish by the tons," she said, "your motivation is pretty strong, especially if your job is to preserve and protect."

Green went straight to the captain of the *Ranger III*, the 165-foot-long ship that ferries park passengers to the island, 73 miles from its home port on Michigan's Upper Peninsula. Worried that the ferry might suck the rapidly spreading virus into its ballast tanks while docked at the mainland, she asked if there were any way to disinfect that ballast before it was released into park waters. The captain said no.

"What happens," Green replied, "if I tell you that you can't move this ship unless you kill everything in your ballast tanks?"

That's when the brainstorming started. Green's goal was to try to figure out how to make the *Ranger III* safe to sail—not in years or even months, but in a matter of days. She sat down with the captain, the ship's engineer and David Hand, the department chairman of civil and environmental engineering at Michigan Technological University. Hand had worked on water purification systems for the International Space Station that can turn sweat and urine into tap water.

"This," Hand told the group of the ballast problem, "is not rocket science."

Two weeks later, Isle Royale's passenger ship had a crude ballast treatment system that used chlorine to fry viruses and other life lurking in its 37,000-gallon ballast tanks, and then vitamin C to neutralize the poison so the water could be harmlessly discharged into the lake. Green didn't stop there. She leveraged her authority as protector of Isle Royale to block all freighter ballast discharges within a 4.5-mile radius of the island, which happened to cover shipping lanes used by freighters sailing to and from the Canadian port of Thunder Bay.

The Park Service has since installed a permanent ballast treatment system on the *Ranger III* that uses filtration and UV light, a first for the Great Lakes. Although the Isle Royale boat is almost toy-sized compared with the freighters that ply the Great Lakes, Green contends the relatively simple chlorine treatment could be scaled up to the biggest boats on the lakes as an emergency line of defense that would be far stouter than the saltwater flushing—the only protection for the lakes until ballast treatment systems are required for all overseas ships, which likely won't be until 2021 at the earliest.

The Seaway operators and ship owners have long acknowledged the ballast problem and are quick to note they support the EPA rules requiring ballast treatment systems. Yet the shipping industry has in the past sided with the EPA in its legal fight to exempt ballast water from Clean Water Act requirements. Shipping advocates have also been dubious the industry can meet the nearly decade-long EPA deadline set in 2013 to require ballast treatment systems on all ships sailing in from overseas ports.

"I'd use the term 'ambitious,'" said U.S. Seaway boss Craig Middle-brook. "Is it doable? We're going to see."

THERE ARE, NOT SURPRISINGLY, OTHER PATHWAYS FOR SPECIES to invade the Great Lakes. Fishermen from other watersheds can dump their bait buckets. Aquarium owners can dump their pets. Anglers hoping to "improve" their fishing prospects can intentionally plant exotic fish. But contaminated ballast water from overseas vessels has been, by far, the dominant pathway for Great Lakes invasions since the Seaway opened.

Contaminated ballast water is also not just a Great Lakes problem. It has been linked to a cholera outbreak that killed more than 10,000 people in South America two decades ago. It is why Chinese mitten crabs and Asian clams are ravaging what's left of the native species in the heavily invaded San Francisco Bay and a cantaloupe-sized snail from Asia called the *veined rapa* whelk is creeping across the bottom of the Chesapeake Bay. But guarding the Atlantic and Pacific coasts from these invaders is a far more daunting task, both because of the volume of overseas freighter traffic servicing the Pacific, Atlantic and Gulf coasts and their geographical vastness.

The Great Lakes themselves are wrapped by some ten thousands of miles of shoreline, every mile of which is vulnerable to the biological mischief ferried in by overseas ships. But unlike on the Atlantic, Gulf or Pacific Coasts, there is, literally, a door through which every foreign Seaway ship must pass before arriving at the dozens of ports rimming the Great Lakes' shores: the St. Lambert Lock at Montreal. Every ship—and every one of the Great Lakes invaders it may be carrying—has to squeeze through this 80-foot-wide pinch point. Stop the overseas ships, known regionally as "salties," and you stop their ballast invasions. "Off-load the cargo in Nova Scotia and ship it down through rail," an exasperated former Chicago Mayor Richard Daley once told me. "That will protect the Great Lakes forever. That will protect local and state governments from spending hundreds of millions of dollars." He is not alone.

Conservationists agree this low-tech solution for the Great Lakes could prove far cheaper than installing on each ship ballast treatment systems that could cost well over a million dollars.

"The solution could be simple," said Todd Ambs, the former chief for the Wisconsin Department of Natural Resources' water division who now works for a coalition of conservation organizations. "You could force the salties to offload their cargo at a point of entry into the lakes. Then have the governments compensate the shipping industry for it. I don't know why we can't have that conversation to see if it's feasible."

That point of entry—the St. Lambert Lock—is owned by Canada, which has made it clear that it is not about to shut it to overseas freighters anytime soon. The Canadian government, in fact, protested vigorously when the state of New York, whose waters all Great Lakes-bound Seaway vessels must also pass through, tried to pass its own ballast water regulations in 2010 that were far tougher than those eventually adopted by the EPA. New York ended up backing down.

The Seaway's Middlebrook acknowledged that locking out the salties may appeal to people outside the shipping industry, but said that is because they don't understand the complexity of the problem. "I know there is clarity to the idea, to achieve one goal, perhaps," he said. "But we're in such a complex interrelated system that the challenges are daunting."

Yet it may be the Seaway and the Great Lakes' overseas shipping industry that misunderstand the problem's complexity, or lack thereof. Nobody disputes that Great Lakes shipping is a big, critically important business. But the vast majority of the cargo moved on the Great Lakes and Seaway—things like salt, iron ore, coal and cement—is carried by U.S. and Canadian domestic freighters that tote those bulk materials from one Canadian or U.S. port to another. Overseas freighters in a normal year account for 5 percent or less of the overall tonnage moved annually on the Great Lakes and Seaway. These ships typically carry inbound shipments of foreign steel and outbound loads of grain, which in 2011 accounted for less than 2 percent of total grain exports from the United States and Canada.

And the total tonnage carried annually by salties sailing into and out of the Great Lakes has been slipping for decades. It is now roughly equal to what could be hauled by a single daily inbound and outbound train from the East Coast. If that train delivered as much ecological havoc as the salties have, it is unlikely the public would still allow it to be running down the tracks. Gary Fahnenstiel, an ecologist retired from the National Oceanic and Atmospheric Administration who has spent much of his career studying the mussel-triggered collapse of the bottom of the Great Lakes food web, argues for a similarly tough approach to the Seaway's overseas freighters.

"It's a reasonable alternative, absolutely," Fahnenstiel said. "The shipping industry may not view it that way, but I guarantee you the people that love the Great Lakes view it as a reasonable alternative."

But what might it cost?

In 2005 two Michigan logistics experts took the first crack at putting a price tag on bringing in the Seaway's overseas cargo into the region by other means. The figure they came up with was $55 million annually. That's what it would cost to transfer the salties' cargo from a coastal port to trucks, rail or regional boats. The study's peer reviewers said if anything the $55 million was an overestimate. Moreover, that figure was derived at a time when it was assumed more than 12 million tons of overseas cargo floated annually on the Seaway—almost double the tonnage in some recent years.

What about the other side of the ledger, the economic cost of Seaway invasions? The overall toll just to municipalities and power companies trying to keep their pipes mussel-free over the last quarter century tops $1.5 billion. This figure is "conservative" and does not give a complete picture of the problem, Chuck O'Neill, director of the National Aquatic Nuisance Species Clearinghouse at Cornell University, told me. That $1.5 billion figure, tallied in 2008, is almost certainly lower than the actual costs because O'Neill said it does not include impacts to manufacturers who typically declined to participate out of fear it would affect their stock prices.

And in terms of damage to fisheries and other recreational activities, the dollar toll for the ecological unraveling of the lakes due to ballast invasions was pegged at $200 million annually, or $2 billion per decade, in a 2008 University of Notre Dame study—a number the study authors predicted would grow as new invasive species are discovered.

The question now is: How will the public respond once the next new invader turns up? Cleveland's industrially fouled Cuyahoga River, after all, burned over and over throughout the 19th and 20th centuries. As late as the 1950s the flames on the oily surface of the water were still viewed as business as usual. But eventually the public had enough, and when the river was set ablaze in 1969, it turned a nation livid and led to passage of the Clean Water Act.

Isle Royale's Green predicts similar fury with the shipping industry when the next ballast invader turns up. "The industry has had this grace period to find solutions," she told me. "The grace period they have been given will hit the fan when they find the next one."

She told me this on a raw, rainy day in 2014 at her park headquarters not far from the shoreline of Lake Superior, and my notes were a smudgy mess. I couldn't tell from my scribble if she said "if" or "when" a new invasion happens. So I called her back a couple of weeks later to get clarification. She chuckled ruefully.

"No," she told me. "I said 'when.' Definitely 'when.'"

PART TWO

THE
BACK
DOOR

Chapter 5

CONTINENTAL UNDIVIDE

ASIAN CARP AND CHICAGO'S BACKWARDS RIVER

The words "continental divide" conjure images of a cloud-scraping mountain crest stretching like a spine down the middle of North America that forces raindrops and melted snow to flow one way or the other—down a western slope and out to the Pacific Ocean or down the other side toward the Atlantic. But a similarly important continental divide runs right through the western edge of downtown Chicago—the one that splits the Mississippi River basin from the waters of the Great Lakes.

The Mississippi River drainage basin covers about 40 percent of the continental United States, stretching from Montana to New York to Texas. This means that the streams, creeks and rivers in the 1.2 million square-mile expanse that is roughly the size of India flow downhill into the Mississippi River and out to the Gulf of Mexico. The Great Lakes basin, which straddles the U.S. and Canadian border, spans nearly 300,000 square miles. Water that falls inside this basin is ultimately gathered by the St. Lawrence River that flows out to the North Atlantic.

These grand watersheds, of course, are not separated by a mountain range. In most places the divide, which runs about 1,500 miles from eastern Minnesota to western New York, is a gently sloping hill. In some places it is merely an imperceptible bump in the landscape. And in Chicago it was, historically, nothing but a marshy area feeding two rivers flowing in opposite directions. The west side of the Chicago divide flowed into the Mississippi-bound Des Plaines River. Its eastern side flowed into the tiny Chicago River, which trickled into Lake Michigan.

Or at least that's how things used to work.

The first European explorers who came upon this sag in the divide in 1673 must have been shocked as they paddled their way back from an expedition down the Mississippi River. Father Jacques Marquette and Louis Joliet planned to return to their base near the northern tip of Lake Michigan along their outbound route, which required an arduous portage across the basin divide in what is now central Wisconsin, but the Native Americans they met on the return trip steered them to this shortcut at the southern end of Lake Michigan. Here the explorers only had to tug their canoes across a marshy area less than two miles wide. Marquette and Joliet knew immediately this soft spot in the divide was among the most strategic points on the continent: if it were someday breached with a canal, it could open a nonstop navigation corridor between Lake Erie and the Gulf of Mexico, and that meant opening the door for trade and supplies to fuel settlement of the middle of the continent.

It took almost two hundred years, but the Chicago divide was finally destroyed in the mid-1800s with a relatively crude navigation channel. This canal removed the last plug in a continental navigation network that finally allowed for goods and people to float up the Erie Canal from the East Coast, across the Great Lakes and then down the Mississippi River and into the Gulf of Mexico.

How important was this 6-foot-deep ditch to the development of Chicago? Its population was less than 5,000 in 1840, at the time the canal was under construction. Just over a decade after the canal opened in 1848, the city's population burst to more than 100,000, and it nearly tripled again in the following decade. Barges pulled by mule across the divide were laden with so much cargo—grain, lumber, livestock and foodstuffs like fruit, sugar, salt, molasses and whiskey—that Chicago, once a swamp in the middle of the continent, had become the nation's busiest port by 1869.

The economic benefits of a canal that cost $6.5 million to construct were astounding, but the true cost of Chicago destroying its continental divide—the price paid for essentially re-plumbing half the continental United States—would not be known for another 150 years. As was the case with the St. Lawrence Seaway and the earlier canals linking the Great Lakes to the East Coast, opening this Great Lakes' "back door" to the waters of the Gulf of Mexico unleashed not only the intended stream of commerce. It also, eventually, let loose an unintended torrent of ecological trouble nobody pondered when the first shovel took the first scoop of earth out of the divide on July 4, 1836.

As in the great Greek myth, the problems started with a box. A box full of fish.

ON A BRISK NOVEMBER DAY IN 1963, A STATION WAGON PULLED up to the brown brick federal research lab in eastern Arkansas loaded with a radical new weed killer. In the wake of the publication of Rachel Carson's *Silent Spring* the previous fall, there was an increasing awareness of the potential perils of all the herbicides and pesticides flowing in our rivers, across our croplands and orchards, down our grocery aisles, onto our dinner tables, into our bloodstreams. Poisoning rivers to get rid of nuisance fish was particularly in vogue at the time, including the Russian River in northern California and Utah's Green River, and a clamor was growing for a smarter, gentler approach to combating unwanted creatures and vegetation. So researchers at the U.S. Department of Interior's Fish Farming Experimental Laboratory, located in the heart of Arkansas catfish farming country, were taking delivery of what they hoped would be a new generation of nontoxic aquatic weed-control agents.

The station wagon's tailgate was dropped and three cardboard boxes, each with two white arrows pointing up, were hauled through the lab doors. The label on the boxes from Malaysia told the handlers that this was not just another toxic chemical compound whipped up in a lab. It read: "Live Fish." The boxes contained dozens of juvenile grass carp, a species native to Asia and famous for taking to forests of seaweed like locusts to crops. The idea at the research lab was to deploy these fish instead of chemicals across the South to clean fish farm ponds as well as weed-choked rivers and irrigation ditches.

"When they did this, this was right. This was the thing to do," said Andrew Mitchell, a recently retired biologist from the federal research lab. "It was one fish to do one job—keep chemicals out of the environment."

That station wagon's payload was the first documented shipment to the United States of a group of fish collectively known as Asian carp. Within a decade of the grass carp's arrival, an Arkansas fish farmer seeking his own batch of the exotic weed-eating fish accidentally imported the three other Asian carp species: black, bighead and silver carp. He didn't know what to do with these other types of Asian carp because they weren't weed-eating machines. Silver and bighead carp are filter feeders that strip plankton and other floating nutrients from the waters in which they swim. Black carp live off mollusks.

The fish farmer did what he thought was the right thing. He turned his exotic fish over to the government. State fishery workers could have destroyed these Adams and Eves. Instead they decided to try to get the novel brood to reproduce—just for kicks, apparently.

"We had this little agreement that if we learned how to spawn them, that he got some of the stock back," former Arkansas Game and Fish Director Scott Henderson told me. "It was all cordial. We were interested in doing some research to see what they were, and I guess at the time, getting them out of the public."

The fish farmer gave Henderson's department 22 adult silver carp, 20 adult black carp and 18 adult bighead carp and, like cavemen trying to spark a fire, the state hatchery workers tried feverishly to get their brood of exotic fish to breed. They had little luck because, it turns out, raising Asian carp in hatcheries is an absurdly intricate procedure that requires precise timing and water conditions, as well as injections of crushed fish pituitary glands and human hormones harvested from the urine of pregnant women.

The hatchery workers killed all the black carp before they could reproduce a single fish, but they had better luck with the bighead and silver carp when they turned to S. Y. Lin, a professor at National Taiwan University who had moved to Washington, D.C., as an employee of the United Nations Food and Agriculture Organization. In his three weeks in Arkansas back in the early 1970s, Lin took two 12-pound silver carp and produced nearly one million eyelash-sized silver carp fry. The one 15-pound bighead carp Lin worked with produced 20,000 little bigheads.

And the Arkansas fish biologists had their fire.

Not long after they had cracked the breeding problem, the Arkansas Game and Fish Commission agreed to send some of their fish to other aquaculture research facilities, including Auburn University. They also entered into a contract with the U.S. Environmental Protection Agency to employ the curious carp in sewage treatment experiments. Former Arkansas Game and Fish chairman-turned-fish-farmer Mike Freeze explained to me the rationale behind the bizarre experiments while bouncing around in his pickup along the levees containing his crops of catfish. Arkansas waterways in the 1970s, Freeze explained, were like waterways everywhere else in the country— ridiculously filthy, in part because small communities didn't have adequate sewage treatment systems.

So Arkansas turned to the filter-feeding carp in an elegant, if a bit repugnant, plan to clean up its waters. Phase One was to plant bigheads and silvers in experimental sewage lagoons and let them convert the decaying human waste into fish flesh. Phase Two would be to sell those fish as food to fund small cities' sewage treatment costs. The fish, particularly bighead carp, are favored in Asia for their flaky, mild flesh.

"I remember we sent sample after sample [of fish] from the sewage ponds to Baylor University to make sure they didn't have any viruses or things like that," said Freeze, who was an hourly worker at the state hatchery during the time of the carp experiments. It didn't take long for word of Arkansas' feces-to-flesh business plan to spread, and the federal Food and Drug Administration soon stepped in. "They had a standing policy," said Freeze, "that it was not legal to take these fish out of sewage ponds and sell them for human consumption."

The experiments stopped when federal funding dried up. Some fish were destroyed. Others were simply set free. Freeze remembers containment screens swinging open and gates being lifted to drain hatchery ponds—and their inhabitants—into Arkansas ditches and streams. It all seemed innocuous at the time—the fish were so difficult to breed under even precise hatchery conditions that nobody thought there was any chance that the carp would be able to breed on their own in the wild. This was a blunder of the highest order. Soon baby bighead and silver carp started turning up in rivers and streams across the South, and the swarming fish have been migrating north ever since.

The problem is bighead and silver carp don't just invade ecosystems. They conquer them. They don't gobble up their competition. They starve it out by stripping away the plankton upon which every other fish species directly or indirectly depends. Bighead carp can grow larger than 100 pounds and each day consume up to 20 pounds of plankton. Bighead and silver carp have so squeezed aside native species that the Asian carp biomass in some stretches of rivers in the Mississippi basin is thought to be more than 90 percent—the same dire situation that an alewife-plagued Lake Michigan suffered in the 1960s.

A fishing boat, brimming with bighead carp.

It's not just native fish species that suffer where these fish succeed. Silver carp, which are slightly smaller than bighead carp, have become YouTube sensations because of their penchant for rocketing out of the water like piscine missiles when agitated by the whir of a boat motor. One such video starts with a *Star Wars*-style opening. Text scrolls in front of a globe spinning in space, and an arcing red line shows the invasive fish's journey from Asia, across the Pacific Ocean to the Illinois River town of Peoria. What follows is some four minutes of young men water skiing while wielding swords and wearing football helmets bristling with giant nails. Fish-impaling spikes poke similarly from shin guards. One water skier protects his torso by wearing a bum-in-a-barrel-style metal garbage can. The water skiers swat, smack and hack at the hundreds of fish bursting from the water like popcorn.

All joking and ridiculously dangerous stunting aside, water skiing and jet boating have become impossible in some places, particularly on the Illinois River where things are so bad that the town of Bath hosts an annual "Redneck Fishing Tournament." It is the one time of year in this southern Illinois hamlet you will find the river packed with motorboats. They are piloted by beer-swigging, helmet-wearing anglers who try to catch as many silver carp as possible in three hours—not with hooks and lines, but by waving fishing nets in the air.

The year I attended the tournament I saw several people leave the water after being battered and bloodied by the jumping fish that hit hard as a fist. I saw one man, not smart enough to wear a helmet, spit out a tooth. The tourney's 78 registered boats that day landed 1,840 fish, all of which were buried with a backhoe on the riverbank. The bar owner who hosted the tournament assured me the mass grave didn't even dent the population swarming in the river, but insisted that was not the point of the event. She hosts it to provide locals a bit of zany, late-summer fun and, more importantly, she said it serves as a warning to people across the country of what could be coming to their own lakes and rivers. She had a particular set of lakes in mind.

"If these things get into the Great Lakes," she told me, "you are in trouble."

The word trouble doesn't really capture what is at stake, both environmentally and economically, if the oversized fish succeed in what has so far seemed like their inevitable push to colonize the Great Lakes, the biggest home they could ever hope to find, and one that still sustains a multibillion dollar commercial and recreational fishery.

While silver carp make the headlines for their leaping ability, bighead carp lurk largely out of the public's consciousness. Not so for commercial fisherman Orion Briney, who more than a decade ago figured out how to eke out a living by catching bigheads on the Illinois River and selling them to a wholesaler who guts, ices and ships them by the refrigerated container-full back to China.

Briney can catch 15,000 pounds of bigheads in his nets. Not in one day. In 25 minutes. Here is a little perspective on that number: Wisconsin's quota for commercial perch fishing on all the state waters of Lake Michigan in some past years has been about 20,000 pounds. That's not a per-day limit. That's the limit for an entire year.

I went out one steamy summer day with Briney and was left gobsmacked (and silver carp–smacked) by what had become of the river since the invaders had swarmed in just a few years earlier. Briney fishes cowboy-style, using his boat to herd his quarry. "See that big wave?" he asked me as we roared downstream at dawn. I could see only a patch of choppy black water. "I'll bet there is 400,000 to 500,000 pounds in there!" He arced his boat toward the fish and then swooped down behind them, chasing the thrashing mass into his nets. Briney had no interest in the silver carp flying about the boat—and his head, which he deftly shielded with his Popeye-sized forearms. He was angling only for bigheads, a tastier and less bony fish that has a small market in the United States and, because bigheads are typically sold live in Asia, has only limited appeal abroad.

It took him less than a half hour to round up more than 13,000 pounds of fish—and another 3½ hours to pluck out the bigheads, one by one, from 800 yards of net. The biggest weighed nearly 40 pounds, monsters that lurked so low in the murky water they're rarely seen by the pontoon boaters and recreational anglers that still dare to venture out on the river. Seeing one up close is unnerving. The fish have mouths big enough to gobble softballs whole and their eyes are so low on their head that they appear to swim upside down.

"Most people don't even know them's here," Briney said as he piled the writhing bigheads on the boat floor. "They just see the silvers jumping." Does he see the carp finding a future in the Great Lakes basin next door? "They'll thrive. There's plenty of food," he said. "They'll love it." Then he asked me: "The lakes are, what, 20, 30 feet deep?" No. The Great Lakes are hundreds of feet deep.

"Good Lord!" he groaned. "By the time [anyone] knew they had a problem, it'd be too late."

Biologists remain dubious about whether bigheads and silvers could thrive similarly in the Great Lakes' open waters, which are relatively sterile compared to the soupy rivers lacing the Mississippi basin. But it may be another matter altogether for the lakes' algae-rich bays and harbors, and the rivers that feed them—which also happen to be the places where most people boat, jet ski and fish. Inland waters connected to the lakes are similarly threatened. The financial impact of an infestation of these areas alone could be staggering. The eight Great Lakes states alone are home to some four million recreational boats, about a third of the United States' total.

Arkansas' Freeze knows these numbers and he has lamented from afar the bighead and silver carp's inexorable march north. He told me he believes that after his crew let their fish go free there were subsequent escapes of bighead and silver carp from research facilities and fish farms (captive-raised bighead carp eventually became a common cash crop for fish farmers, though it is now illegal to transport them live across state lines.) He also provided me documentation showing that the federal government later helped fish farmers reimport black carp to control pesky snails in aquaculture ponds. Not surprisingly, black carp are now also swimming free in rivers down South.

But he acknowledges that the bighead and silver carp that got loose on his watch almost surely were among the first to get into the wild. He's not proud about his role in what is shaping up as a billion-dollar blunder. But he doesn't hide from it, either. "I'm old enough and big enough," he said, "to say that there are a lot of things in my life that I'd go back and change."

Had Chicago just stuck with its first continental divide–busting canal, the carp could probably have been blocked from entering the Great Lakes with little more than a caravan of dump trucks loaded with sand and gravel. But, as was the case with the earlier canals that opened the Great Lakes to the Atlantic Coast, big enough was never big enough for long.

IN THE OPENING STANZA OF HIS 1916 ODE TO CHICAGO, POET CARL Sandburg playfully refers to his City of the Big Shoulders as Hog Butcher for the World and Player with Railroads. He might also have added Conveyor of Crap, because less than two decades earlier Chicago built what is essentially a continental-sized commode, turning Lake Michigan into the world's largest toilet tank, and the Gulf of Mexico into its toilet bowl. It was a matter of life and death for a mushrooming city that sent its sewage into Lake Michigan, from which it also takes its drinking water.

In the 1890s, in order to protect that drinking water, engineers began work on the Chicago Sanitary and Ship Canal. The city's motives for the canal are made clear in its name; when it opened in 1900, its primary job was to flush the city's waste across the continental divide and into the Mississippi River basin. It also, conveniently, doubled as a massive expansion of the original barge canal linking Lake Michigan to the Gulf of Mexico.

The rail-straight, 25-foot-deep canal that is as wide as a football field does a most remarkable thing. It reverses the flow of Chicago's namesake river, which was in its natural state a shallow, slow-flowing dribble feeding Lake Michigan. Its headwaters were just several miles west of downtown at the bog Marquette and Joliet first came across in

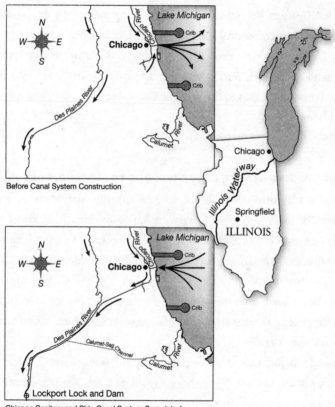

The Chicago Sanitary and Ship Canal destroyed the natural hydrological divide between Lake Michigan and the Mississippi River basin, reversing the flow of the Chicago River.

the late 1600s. The Sanitary and Ship Canal, which is lower than Lake Michigan, connects to this river. This pulls the formerly lake-bound river's flow backwards. Instead of the river feeding Lake Michigan, the lake now feeds the river. The river then flows into the canal, the canal flows through the continental divide and into the Des Plaines River, which flows into the Illinois River, which flows into the Gulf-bound Mississippi River.

The canal did exactly what Chicagoans hoped: it preserved the source of their drinking water. It also brought immense controversy in Mississippi River-drinking cities like St. Louis, which did not relish the notion of consuming Chicago's sewage, even if it was diluted by the 357-mile trip down from Chicago.

In early January 1900, just as its construction neared completion, the state of Missouri petitioned the U.S. Supreme Court to block opening the canal's connection to the Des Plaines River. That prompted Chicago leaders to sneak out of town on a train in the early hours of January 17 to open the downstream canal gates before the court could stop them. There was little pomp in a ceremony the *New York Times* characterized as one conducted with "undignified haste," at the conclusion of which a pale green tongue of Lake Michigan water crashed into the Mississippi basin. Chicago's raw sewage wasn't far behind, and it didn't take long for the fouled downtown stretch of river to be flushed of its accumulated excrement—a city-sized enema. "Water in the Chicago River Now Resembles Liquid," a *Times* headline deadpanned in the days after Lake Michigan water first entered the upper portion of the canal. The Mississippi River basin and the Great Lakes have been unnaturally connected by a canal wider than a Los Angeles highway ever since.

In making its case to the Supreme Court, Missouri reported that the annual number of typhoid fever cases in St. Louis had approximately doubled in the four years after the canal opened, compared with the prior four years. Even so, Missouri had a hard time convincing the justices of the Supreme Court that Chicago toilets were the problem. It was the dawn of the 20th century, and the science of microbiology was in its infancy. The plaintiffs contended that the typhoid bacillus could survive the 8 to 18 days it took for Lake Michigan water to flow to St. Louis. The defense argued it could not. And the justices were left dubious about the danger posed by an invisible menace.

"There is nothing which can be detected by the unassisted senses— no visible increase of filth, no new smell," Justice Oliver Wendell Holmes Jr. wrote in the court's majority opinion, finally rendered in 1906. "On the contrary, it is proved that the great volume of pure water from Lake Michigan which is mixed with the sewage at the start, has improved the Illinois river in these respects to a noticeable extent. Formerly it was sluggish and ill smelling. Now it is a comparatively clear stream."

Today it is clear that dangerous—even deadly—microbes can lurk in the purest-looking glass of water. But more than 100 years ago it was a different story. "The plaintiff's case depends on an inference of the unseen," Holmes concluded. This would not be the last time the federal government turned a blind eye to invisible trouble lurking in the canal.

St. Louis's typhoid fever troubles evaporated with advances in water treatment in subsequent decades, and Illinois' Great Lakes neighbors long ago learned to live with a slightly diminished Lake Michigan because the canal was built to siphon away up to six billion gallons of lake water per day (the subject of another Supreme Court case filed by Wisconsin in the 1920s that also failed to end Chicago's Great Lakes water grab). But by the late 20th century, the larger, unforeseen costs of tinkering with the hydrology of a continent were coming into focus, and they had nothing to do with water levels on Lake Michigan or downstream cases of diarrhea. It became apparent that Chicago had accidentally built a superhighway for Great Lakes invasive species to fan out across North America.

As we've seen, ships sailing up the St. Lawrence Seaway have dumped dozens of exotic critters into the Great Lakes since it opened more than a half century ago. The Chicago canal system turned this regional problem into an ever-growing national one as invasive species ride its waters out of the Great Lakes and into the basin that connects the waters of 31 states.

Scientists have identified 39 invasive species poised to ride the Chicago canal into or out of the Great Lakes, including a fish-killing virus in Lake Michigan today that could ravage the South's catfish farming industry as well as five species of nuisance fish, including the sea lamprey. Threatening from the other direction, beyond the Asian carp, is the razor-toothed snakehead, which can breathe air and slither short distances over land and is now swimming loose in the Mississippi basin.

This problem was recognized nearly two decades ago, when Congress, alerted to the idea that Great Lakes invaders were primed to make their way out of the lakes and into the Mississippi basin, authorized construction of an experimental electrical barrier on the canal, about 35 miles downstream from Chicago's Lake Michigan shoreline.

Underwater electric barriers to block migrating fish had been successful on streams and irrigation canals in the West, but never on a waterway as big as Chicago's, and never on one that is a major navigation corridor plied by thousands of barges annually. After six years of designing, building and debugging, the Army Corps finally turned on the $1.5 million contraption in 2002. But by then the species of Great Lakes invasive fish that scientists had feared would use the canal to escape into the Mississippi basin had already done so. So the barrier was repurposed as a device to halt the migration of Asian carp into the lakes from the other direction.

The barrier was always intended to be just a Band-Aid until a more permanent solution could be devised. But soon after it was activated, Congress became convinced it was worth funding a giant version, one built to last decades and operate at four times the strength of the original. Construction started in 2004 and was finished in 2006. But then the government refused to turn it on, largely because of the U.S. Coast Guard's fear that its electrical current would cause sparks to fly between canal barges, some of which carry petroleum and other flammable materials. The Coast Guard's primary mission, after all, is to protect the barge workers and recreational boaters floating up and down the canal. It's not to worry about what type of fish are swimming in the Great Lakes, even if it's a type of fish that some fret could utterly upend the world's largest freshwater system. As the former chairman of the commission that helps guide U.S. and Canadian management of the Great Lakes has put it: if the fight to stop the giant fish from colonizing the lakes fails, well, "it is just a matter of time before we end up with a carp pond."

BARRIER SAFETY TESTS DRAGGED ON INTO EARLY 2008 WHEN Army Corps General John Peabody arrived to take charge of the carp fight. Peabody is a 1980 graduate of West Point who has done tours in the Pacific, Panama, Somalia and the Middle East, where he led 3,000 engineers into Iraq during the 2003 attack on Baghdad. In his three-plus decades in uniform, Peabody has picked up a graduate degree in public administration from Harvard, studied as an Olmsted Scholar in Mexico City and earned his master parachutist badge. He has received a Bronze Star for valor as well as a Purple Heart. The general limps on a metal hip. He has a penchant for quoting war movie dialogue and commands most any room he walks into with a no-nonsense demeanor that borders on brusque.

But he was also once a little boy who relished hot summer days at Lake Erie's Nickel Plate Park beach in Huron, Ohio. "The night before, we'd get picnic baskets, beach balls, all the rest, in our station wagon—

this was the '60s . . . and my brothers and I would get all excited: 'Daddy, Daddy! Mommy's taking us to the lake!'" the general told me, as he knocked back a late-afternoon coffee at the massive conference table in his office in downtown Vicksburg, Mississippi. "And my cynical dad would say: 'Your mother is taking you to the biggest cesspool in northern Ohio.'"

Despite the stench of all the rotting alewife carcasses littering the polluted beach when he was a child—one of which Peabody vividly recalls gouging his foot—the Great Lakes stole a soft spot in the heart of the hard-nosed general that motivated him when he took over the carp fight. "There was an opportunity for us to prevent a really bad thing from happening—a calamity, a crisis, whatever word you want to put to that."

In the months after he took over the job, Peabody turned on the new electric barrier that the Coast Guard had kept mothballed. As a safety compromise with his sister agency, Peabody ordered the new barrier, which was designed to run as high as four volts per inch, to run only at the same voltage as the nearby demonstration barrier: one volt per inch. One volt does not provide a strong enough jolt to stop juvenile fish, which, because of their size, are more immune to electrical pulses. But one volt was a level the barge industry could live with and, at the time, no Asian carp had been detected near the barrier.

"If the fish weren't close enough to be a threat, it didn't seem prudent at that time to raise the operating parameters," Peabody told me. That was a big if, and the general knew it. Biologists tracking the pace of the Asian carp's migration up the Mississippi and Illinois Rivers believed that by 2008 the fish should have long ago been probing the barrier, even though crews using nets and electroshocking devices continued to turn up zero evidence of fish in the area. Everyone involved knew that finding the first few fish at the leading edge of the invasion would be exceedingly difficult because the fish have an uncanny ability to avoid nets. Shocking the water to stun the fish so they float to the surface doesn't work either, because Asian carp can often lurk too deep to be stunned.

Even though the nets used to hunt the carp near the barrier came up empty, Peabody still smelled trouble in the canal. "Our lack of information was so great," he told me. "I felt we had to take whatever we could and apply it as quickly as possible to try and get more information."

The University of Notre Dame's David Lodge had what he was looking for.

Lodge is one of the country's leading invasive species experts with a reputation as a scientist who isn't content to see his work go dusty on library shelves. The bespectacled 60-something with a younger man's shock of black hair still has a bit of a Southern drawl from his boyhood in Alabama, yet he enunciates his words in such a precise manner it is easy to picture him as the budding naturalist he remembers being as a child.

"I was one of those kids who just was fascinated with nature from the get-go," Lodge told in an interview in his office at Notre Dame's "Innovation Park," a gleaming new building where fingerprint scans open locked doors. "I spent all my idle moments turning over rocks in streams and swimming, snorkeling, fishing, catching frogs and snakes and turtles and whatever else I could catch . . . I spent my free time inside reading field guides. That's not your average teenager activity. But I was more happy doing that than playing baseball."

Lodge thought about studying history when he got to college, "But in the end I think it was pretty clear to others, even if it wasn't always clear to me, that I just loved biology." That love carried him to Oxford University as a Rhodes Scholar. He went on to serve as chairman of President Bill Clinton's Invasive Species Advisory Council and to create Notre Dame's Environmental Change Initiative, a team of university researchers that aims to inform public policy decisions on hot-button environmental issues such as invasive species and climate change. Crossing the line from pure academic research into public policy wasn't something he did lightly because at the beginning of his career it wasn't even considered acceptable.

"When you wrote proposals to get research to support your work, you didn't couch them in terms of what problems you were going to solve in the world," he said. "You couched them in terms of intellectual excitement and new ideas." But that line has blurred in recent years, and today Lodge's work is often at the center of some of the Great Lakes region's prickliest ecological and political debates. He's done research to predict which species are most likely to invade the Great Lakes if ships are allowed to continue discharging contaminated ballast water; he's helped put a price tag on the annual cost of invasive species to the Great Lakes (an estimated $200 million); he's done work predicting which freshwater fish species are most likely to go extinct because of climate change.

Straddling the worlds of politics and science has never been a comfortable exercise for Lodge. Messengers, after all, sometimes get vilified for delivering grim news. But Lodge came to figure the stress of publicly defending his work in the media and to policy-makers was the price for doing science that mattered. By summer 2009, with Asian carp migrating toward the Great Lakes and federal officials desperate to find someone who could show them exactly how far the fish had advanced toward Lake Michigan, that price was about to explode.

Lodge's skills as an ecologist and his willingness to wade into sticky issues had made him and his colleagues a logical choice a few years earlier when a think tank funded by the Great Lakes states gave his lab a grant to develop a genetic-based test to identify invasive species hitchhiking into the lakes in the ballast water of Seaway freighters. Law enforcement investigators have been using DNA analysis for more than two decades to put criminals behind bars. These genetic fingerprints can be harvested from almost anything the human body sheds—flecks of skin, strings of saliva, drops of semen, strands of hair. From that material, scientists can isolate and identify the molecules that are an individual's DNA, the famous double helix. Each minuscule twisting ladder is made up of billions of rungs built from four types of chemicals, called nucleotides. DNA is such a powerful forensic tool because the order of these billions of rungs, each made up of two interlocking nucleotides, is unique for each individual. Scientists zero in on relatively short sequences of nucleotides on a piece of human DNA to see if the genetic material harvested from a crime scene is a match for DNA taken from a suspect.

But this genetic fingerprinting process also works on the species level; all silver carp, for example, share an identical sequence of nucleotides at various places in their DNA. It wasn't a big leap for the Notre Dame team to realize that DNA testing could be used to find evidence of carp in the canal. This kind of analysis had already been done on a smaller scale by an Italian researcher who used traces of DNA to find American bullfrogs in European ponds. It works because fish and other aquatic life constantly shed cells in mucus, urine and feces. Those cells tend to stay suspended in water, and that means every fish leaves in its wake a genetic trail. That trail can be traced by filtering all the DNA from all the different species that have left behind a piece of themselves in a water sample.

Once that pile of DNA is isolated, lab technicians put it in a test tube and add to it some precisely engineered genetic markers—called primers—that are designed to attach only to the DNA of the targeted species. A concoction with free-floating nucleotides is also added to the mix and then the sample is heated. The heat unravels the DNA helixes of all the species filtered from the original water sample. If any of the targeted species' DNA is present, the primers glom on to each separated helix as the sample cools. That starts a zipper-like reaction in which an enzyme that is added to the sample binds the free-floating nucleotides to each strand of original DNA. Suddenly one piece of DNA has been turned into two. The process is repeated over and over so that even a single piece of DNA can be replicated beyond a billion, to the point the target DNA can actually be seen as a glow under an ultraviolet light when yet another chemical is added. One piece of DNA wouldn't be enough to identify a species in a sample, nor would 100,000. But once you get a billion or beyond, a visible glow emerges.

Now a previously invisible fish is revealed.

IT ALL WORKED BEAUTIFULLY IN THE NOTRE DAME LAB, BUT Lodge's team knew there was a big difference between isolating DNA floating in aquariums and sifting it from a free-flowing river. By early 2009, Lodge's staff was ready to try. At a January meeting in downtown Chicago among researchers guiding operation of the Army Corps' electric fish barrier, one of Lodge's assistants pulled an Army Corps biologist into a quiet corner. He told her he believed they had cracked the problem of filtering and identifying Asian carp DNA from open water and he thought it could be applied on the Chicago canal. The biologist took the idea to her bosses at the Army Corps' regional office in Chicago and got the go-ahead. At that time, Peabody had no idea what his staff or the Notre Dame scientists were up to.

Andy Mahon, an ecologist and geneticist who worked in Lodge's lab back then, remembers the miserable morning a few months later when he and a colleague gave their new fish-hunting tool a whirl on the muddy, spring-swollen Illinois River. They figured if DNA didn't turn up in a place like this, known to be thick with Asian carp, there was no sense trying to detect it where there might be only a handful of the fish. The two spent the morning freezing their hands filling two-liter plastic bottles, the excitement they had felt just weeks earlier in the lab tumbling away as they worked. How could they possibly find mere molecules of fish in all this murky water? Mahon headed back to South Bend with his spirit as chilled as the bones in his fingers. He was alone in his lab testing the samples a few days later when the telltale glow emerged. He dashed down the hallway looking for Lodge and the others.

"Shocked," is how he described their collective reaction.

The team decided to move the testing slowly up the river, toward areas where carp numbers were known to be lower. "We had developed the tool and tested it the best way we could—in the lab, and in the field in a preliminary way," Lodge said. "But to build our own confidence and build the confidence of anybody else, we wanted to start in places where everybody agreed there were fish. So the general strategy was to start south and work our way north [toward the barrier], because the whole idea was to identify where the leading edge of the invasion front was."

When Peabody finally got word of what the Notre Dame scientists were up to, he requested a face-to-face meeting. In the summer of 2009, the general and his staff showed up at Rosie's Family Restaurant just down the road from the electric fish barrier in an industrial corner of suburban southwest Chicago. Peabody and his staff, as usual, arrived in combat dress—camouflaged pants tucked into high-laced boots—to grill one of Lodge's colleagues on what the Notre Dame team was trying to do.

Peabody planted himself at the head of a table with the Notre Dame scientist at his side (Lodge was not there that day; he had a class to teach). The general's staff scattered around, some standing, some sitting at the table. A map was unfolded. Sugar packets were used to represent fish, barriers and boats. It was at times an awkward summit between military men who were demanding clear, yes-or-no answers and a scientist who makes his living in the fuzzy place at the edge of human knowledge.

Lodge's crew knew it was wading into murky waters. For one thing, the specific technique it developed to hunt for Asian carp in rivers hadn't at that point been published in a scientific journal, which meant it had not been independently validated by other scientists. What's more, this type of DNA analysis can indicate nothing about numbers of fish, their precise location (DNA drifts on the current), exactly how long that genetic material has been in the river or even how it might have gotten there. But Peabody was determined to find out if the fish were pressing up against the new barrier. If the general could demonstrate that the giant carp (or even their tiny offspring) had indeed arrived at the barrier, that could justify cranking up its voltage. Peabody heard enough that day to be convinced DNA was the best tool he had to find the fish. He allowed Lodge's team to press on.

The Notre Dame team continued northward in its testing—and continued to turn up evidence of the fish. In September 2009, it reported Asian carp DNA about 10 miles farther upriver than the fish had ever been seen. If the DNA evidence was correct, Asian carp had passed through the last navigation lock before the electric barrier. A lock is a relatively tricky obstacle for a group of fish trying to colonize new waters because a fish has to accompany a boat into the lock chamber and exit with it once the boat is lifted and the lock gates are opened. Think of a cockroach using an elevator to migrate between floors of an office building. It probably will happen eventually, but it involves a certain amount of luck. Then, to establish a breeding population, others have to make the same trip. Then they have to find each other.

The general might not have been pleased with fresh intelligence that at least one fish had apparently breached the last lock before the electric barrier. But at least this new DNA tool seemed to work precisely the way he had hoped. Like a pair of night-vision goggles, it illuminated a previously invisible enemy, and that gave the general a chance to fight back. Peabody doubled the barrier voltage.

The Notre Dame team pressed on with its sampling. Lodge didn't plan to stop until he got to an area of the river where all sampling showed no trace of DNA. "The whole point," he said, "is to go to where we got all zeros, and of course, everybody, including us, was hoping all zeros happened below the barrier."

Everything changed on November 18, 2009, at 7:48 a.m. That's when Lodge sent an email notifying Army Corps officials that water samples beyond the barrier had tested positive for Asian carp. This was not a memo Lodge wanted to write. When it was time to hit the send button, he said, "It made me feel a little sick."

Lodge reckoned a positive sample above the barrier meant at least some Asian carp had somehow gotten through. Like everything else, DNA can't drift upstream. Environmental groups and politicians from neighboring Great Lakes states pounced at the news that the Great Lakes' last line of defense against an Asian carp invasion had apparently failed. They demanded that the Army Corps stop operating two navigation locks near downtown Chicago. The idea was to use the locks as makeshift dams to block the carp's final advance into Lake Michigan. The problem is that closing the gates would not just stop fish, it would also stop the free flow of cargo on the canal, a move that Illinois barge operators said would have disastrous consequences not only for their businesses but for the industries they serve, as well as the fleet of recreational and tour boats in downtown Chicago. But just as it is dubious that closing the St. Lawrence Seaway to overseas traffic would have devastating international economic consequences, the financial impact of shutting the locks (which would affect less than one percent of the total cargo moved through the Chicago region daily) was probably overblown by the barge industry.

Even so, Peabody, who is in the business of moving barges, wasn't about to recommend lock closure, which he argued wouldn't necessarily work because the structures are aged and leaky and the fish might pass through in any event. The locks would also have to be opened if big rains hit because when the water level gets exceptionally high in the Chicago canal, the whole system reverses and the Chicago River temporarily flows backwards into Lake Michigan. Not opening the locks to allow this to happen could inundate downtown Chicago.

In any case, Peabody clearly felt he was already doing enough. He'd cranked up the voltage at the barrier and, just days before he learned that Asian carp DNA had been found beyond it, his agency backed a most radical plan to literally stop dead any migrating fish.

ON THE GRAY MORNING OF DECEMBER 3, 2009, THE CHICAGO SAN-itary and Ship Canal looked like a crime scene. Yellow police tape laced the banks. Roads to the water's edge were blocked by police officers shivering in the cold, unable to explain to passersby precisely what had happened. Behind the barricades, a generator thrummed outside a huge government tent with computer workstations and coffee for the 400 federal, state and Canadian fishery workers who had descended on this industrial corner of Chicago from as far away as Quebec. Just outside the tent, the bosses of the operation corralled the cluster of news reporters at the water's edge to tell their story. They were the ones who had killed the canal, they explained. They had poisoned it because they were at war—with Asian carp.

The first hint that the river was dying came when the fish started to float to the surface, their white bellies aglow in the lifting dawn light. One by one they popped into view, the way stars emerge at dusk. Some could only flap their gills as they drifted on the tea-colored current. Others thrashed. All of them—ultimately a constellation of thousands—would be carcasses by the time the winter sun slipped below the horizon.

"I can sum up my comments in six words: it's time to man the barricades," said the U.S. Fish and Wildlife Service's John Rogner. "For nearly 10 years we've watched as two species of introduced Asian carp—the bighead carp and silver carp—have moved up the Mississippi and Illinois Rivers, and now they are here. They are now at the gates to the Great Lakes and our action over the next several days is designed to protect those gates."

The kill had been ordered because the new barrier needed to be briefly shut down for maintenance. No actual Asian carp had been found within 15 miles of the barrier—about 50 miles from Lake Michigan—but DNA tests were telling biologists that fish were in the canal much closer to the barrier. Even though the original, smaller electric barrier would continue to run during the new barrier's shutdown, Peabody and everybody else in the carp fight wanted the canal sterilized, just in case.

"I'm likening this action to chemotherapy," President Barack Obama's handpicked "Great Lakes Czar" Cameron Davis lamented in the days leading up to the morning the government killed the canal. "Nobody wants to go through chemo, but you do it to protect the good cells from being overridden by the bad cells. That's what this is."

By midmorning on the day of the poisoning, Rogner was assuring reporters the fish kill was going precisely as planned. But there was a problem. Among all the dead bass, bullhead, gizzard shad, suckers, common carp and catfish bobbing to the surface, no one had found a single Asian carp. Lodge's DNA samples might have told Peabody that the carp were indeed swimming in the poisoned stretch of canal, but the fish floating to the surface told a different story.

Night fell. In all the barrels of toxic carcasses headed for a landfill—

ultimately the poisoning of the canal would yield about 54,000 pounds of flesh—there was not even one Asian carp.

Finally, at 7 p.m., Rogner again summoned reporters to the canal's edge. They had found what they were looking for—a single, 22-inch bighead carp. Although it was unsettling news for the Great Lakes that the Asian carp—or *an* Asian carp—had indeed arrived in the waters just below the barrier, a wave of relief washed over Lodge as word reached the Notre Dame team. The sight of that one fish, he thought, at least validated DNA as a fish-hunting tool. Not everyone was so convinced.

Less than two weeks after the poisoning, a team of federal scientists arrived in South Bend to inspect Lodge's lab. They scrutinized how water samples were stored, how they were filtered and what steps were taken to ensure Asian carp DNA hadn't somehow contaminated any of the equipment, which could result in false positive results. The Notre Dame team felt a bit like its spouse had hired a private investigator to catch it cheating. The investigators found no such evidence. In fact, they came away impressed, writing in their official report that Lodge's process is "sufficiently reliable and robust in reporting a pattern of detection that should be considered actionable in a management context."

The Notre Dame team took that to mean that if you've got DNA, you've got strong evidence that the fish are there, and you could justify taking action to somehow plug the canal and block others from making the same trip. Still, Peabody wasn't ready to believe that a positive DNA sample in the canal on the lake side of the barrier meant the fish were swimming free into Lake Michigan. The general worried the DNA might have gotten there some other way. Perhaps a barge coming up from the Asian carp–infested waters of the Mississippi basin had somehow picked up Asian carp slime or excrement on its hull. Or maybe the DNA came from the droppings of a migrating bird that had eaten an Asian carp on its way toward Lake Michigan. Or, perhaps, an

Asian carp flopped onto the deck of a barge below the barrier, and then flopped off it once the barge passed through.

Peabody knew he was taking a chance adopting a novel, untested technology in a high-stakes fight, but he felt compelled to find out where the fish were so he could make the smartest decision possible about how high to operate the barrier voltage. That was all he wanted from Lodge's DNA sampling. He never expected DNA to surface beyond the barrier, but he was taken aback by the fuss it caused when it did.

"We got a few hits above the barrier," he told me, "and some people were flat-panicked by that."

But it wasn't just "some people." It was the attorneys general from five Great Lakes states who went to federal court in early 2010 to force the Army Corps and the State of Illinois to shut the navigation locks as a last-ditch effort to halt the advance. In just a matter of weeks, the decade-old saga of a giant, leaping type of carp slowly migrating north toward the Great Lakes had gone from a regional news story framed as a quirky situation on a curious river to national news about how two species of carp were tearing apart a region.

"Everybody got into this expecting we wouldn't have any positive samples above the barrier," said Lindsay Chadderton, an invasive species expert from the Nature Conservancy who helped the Lodge team develop and deploy the DNA testing. "The reality is if we had only found positives below the electric barrier . . . none of this controversy would have happened. The minute we started finding positives above the barrier, that is when the blame game started. That is when the litigation started with the lock closure, and that is when things started to get testy."

The controversy over how much weight should be put on mere molecules floating in water became A-section news in media outlets as far away and as big as the *New York Times*, which reported that the fight to beat back the fish had so consumed President Obama's Chief of Staff,

Rahm Emanuel, that when he returned to Chicago in 2010 to run for mayor, Emanuel's own staff sent him off with a dead Asian carp.

MOST BIOLOGISTS ACKNOWLEDGE THAT A SMALL NUMBER OF fish advancing beyond the barrier does not mean a self-sustaining Asian carp population in the Great Lakes has arrived, or that one is inevitable. For an invasion to be successful, first the fish have to be sexually mature, then they have to find appropriate spawning areas, then they have to find each other, and then their offspring have to figure out how to survive to adulthood in a new environment and find their own mates. Then the next generation has to have similar success. And so on.

For Peabody, this meant it wasn't time to retreat and move the fight to the locks near Lake Michigan shoreline; it was time to stand his ground at the electric barrier. Meanwhile, Lodge's team continued testing the waters beyond the barrier throughout the spring of 2010, and kept getting more positive DNA hits. Yet fishing expeditions with nets and electroshockers in the same areas continued to turn up zero Asian carp. Desperate to find an actual fish above the barrier and not just its genetic fingerprints, in late May 2010, six months after the first poisoning, the federal government and state of Illinois conducted a second river poisoning just six miles from the Lake Michigan shoreline in an area of water above the barrier that had repeatedly tested positive for Asian carp DNA.

This poisoning claimed another 100,000 pounds of fish representing some 40 species. Not one Asian carp was found. Peabody and the rest of the federal team in the carp fight walked away after the second poisoning more confident that the barrier was holding back the fish, and less confident that Lodge's DNA samples meant fish were in the area.

Peabody and his colleagues with the U.S. Fish and Wildlife Service wanted an actual fish—and not just a lab report—before they would be convinced an actual invasion above the barrier was under way. This is harder than you might expect. Duane Chapman, a United States Geological Survey biologist, once marshaled four boats to chase three radio-tagged Asian carp for two days on a Missouri River tributary using electroshocking gear and commercial fishing nets. The radio tags emitted a signal that told the crew the precise location of the fish, which had been trapped between a set of two nets stretching the entire width and depth of the river. But the water was too deep for the electro-shockers to force the fish to surface, and the carp proved cagey enough to avoid getting snarled in the nets.

"They know what nets are," Chapman said, "and they avoid them."

Finally, on June 23, 2010, a month after the second river poisoning and about a year after the Army Corps first hired Notre Dame to help it find the leading edge of the invasion, the Illinois Department of Natural Resources announced that a 20-pound adult bighead carp had indeed been caught in a fishing net beyond the barrier—just six miles from the Lake Michigan shoreline. "We set out on a fact-finding mission and we have found what we were looking for," Rogner of the U.S. Fish and Wildlife Service proclaimed in a news release.

Lodge again felt a grim sort of vindication and Illinois' neighboring states clamored louder for lock closure, arguing that the fish was, finally, irrefutable evidence the electric barrier was leaking Asian carp. Then a few weeks later Rogner issued another news release that claimed the fish might have been lifted around the barrier by human hands after a lab analysis showed that apparently it had spent at least some of its life in waters below the barrier. He based this theory on a lab analysis of the carp's ear bones, which can sometimes tell researchers where a fish once lived, because different water bodies leave different chemical signatures on the bone. But not in this case—an independent reviewer of the lab analysis warned that no definitive conclusion could be reached about this particular bighead carp's life history. Even so, the news release claimed that the bone analysis "does suggest to us that the fish . . . may have been put there by humans, perhaps as a ritual cultural release or through bait bucket transfer."

The problem, beyond the peer reviewer's caution not to use this fish's ear bones to infer anything about its life history, is that an ear bone doesn't contain any information about *how* a fish moves from one water body to another; it can't reveal if a fish traveled in a tank in the back of a pickup truck or if it made the migration on its own. As for the story about a human planting it, bighead carp were at one time sold live in fish markets around Chicago, and rumors of "cultural releases" of Asian carp by people of Asian ancestry practicing some sort of ritual have swirled around Chicago for years. But when I pressed an Illinois Department of Natural Resources spokesman for any evidence of this actually happening with bighead or silver carp in Chicago, the best he could muster was a link to a Wikipedia article. It stated that animal releases were a common practice during the Ming Dynasty in China.

He didn't offer to send me the Wikipedia article that mentioned the Ming Dynasty ended in 1644.

PEABODY AND LODGE EVENTUALLY STOPPED COMMUNICATING and the two ended up in a downtown Chicago federal courtroom in early September 2010. They were both called as witnesses in the federal lawsuit brought by the neighboring states. In addition to lock closures, the states sued to force the Army Corps to expedite a congressionally ordered study of what it will take to permanently reconstruct the natural barrier between the Great Lakes and the Mississippi basin that the Chicago canal destroyed. When Peabody took the stand, he looked mildly agitated. "Never in my worst nightmares or wildest imagination did I think that a fish would so dominate my time and attention, because you don't think of the Corps of Engineers doing that," he told me later. "You think of us as managing water resource infrastructure and building things and fighting floods and doing that kind of stuff."

During his day in court, Peabody talked about his doubts that positive DNA samples meant the presence of live fish, and he had a fish expert ally in the courtroom in Charlie Wooley, the deputy regional director of the U.S. Fish and Wildlife Service. Wooley testified that the federal government and state of Illinois had gone to extraordinary lengths to try to follow up on the positive DNA samples with evidence of actual fish. They hunted for the fish with nets, electroshockers and barrels of poison. And all they had to show for it was that single bighead carp above the barrier, and a single bighead carp in the waters just below the barrier—the fish killed in the December 2009 poisoning.

"The traditional methods allow us to go out and sample large areas very effectively, very efficiently in a relatively short period of time," Wooley testified. "They are tried and true."

Lodge countered on the witness stand that those traditional fish sampling tools are effective at capturing only 1 percent of a fish population on a river the size of the Chicago canal. That means if small numbers of Asian carp lurked among tens of thousands of fish, odds are they would never turn up in a net or float to the surface if shocked. Asian carp, unlike many other fish species, also have a tendency to sink when poisoned. All this, Lodge has long argued, is precisely what makes DNA surveillance so powerful. But it's only as powerful as the trust you put in it.

"Continuing to try to use the traditional tools to detect the presence of rare or of very sparse populations is like saying: 'You know, even though we've got an MRI machine, I'm going to try and detect your cancer with a physical exam; I don't really understand what this MRI machine is doing, so I'm not going to believe it. I'm going to poke you with my fingers and figure out whether you've got cancer or not,'" he said in later interview.

Lodge was shaking as he ate from a bag of M&M's after his testimony. He looked as if he had been the one on trial. The judge sided with the Army Corps and refused to issue an emergency order to shut the locks.

In early 2014, the Army Corps finally released a 10,000-page plan to permanently plug the canal to re-separate the two watersheds, a project that would require extensive sewage treatment upgrades because much of Chicago's treated waste would once again flow into Lake Michigan. The agency said the project, which would include a transfer station to move barge cargo over the barrier, would take decades to construct and cost as much as $18 billion—a schedule and price tag critics contend is wildly overblown. An engineering study funded by a group representing the Great Lakes region's mayors and governors, in fact, concluded it could be done for as little as $4.25 billion and a barrier could be in place in a matter of years, not decades.

The project remained on hold as of 2016, and its opponents are confident it never will happen. "I've been lobbying 25 years on behalf of industry," said Mark Biel, the executive director of the Chemical Industry Council of Illinois. "I'm pretty good at killing bills and ideas that people come up with, and this one has all the elements you'd need." Biel then listed those elements: the time it would take to accomplish; and the cost; the legal, regulatory and political hurdles tied to sending at least some of Chicago's treated wastewater back into Lake Michigan. "This is not going to happen in my lifetime," the 51-year-old said. "And I don't plan on dying anytime soon."

Peter Annin, a former *Newsweek* correspondent and author who also previously worked with Lodge at Notre Dame, takes a longer view. He expects pressure to patch the breach between the basins will surge every time a new invasive species is found heading for the canal—whether it's swimming toward the Great Lakes or out of them, and into the rest of the continent. "This is not about Asian carp," Annin said. "This is about two artificially connected watersheds that many people argue never should have been connected."

THE DNA EVIDENCE THAT ASIAN CARP ARE SWIMMING IN LAKE Michigan has mounted in recent years. It turned up some 200 miles north of Chicago in the waters of Wisconsin's Door Peninsula in 2013. It turned up in 2014 in an inland river in the state of Michigan, one that flows into Lake Michigan. It also turned up in late 2014 in the Chicago River just about a block from the lakeshore. But these microscopic flecks, this genetic "smoke," is all that researchers have found since the one and only Asian carp was plucked from waters above the barrier in 2010. The lack of fish flesh does not matter to Lodge's colleague Chadderton, who has likened the genetic evidence to that of a murder scene.

"If it was a single fingerprint on the murder weapon and in the house where the crime occurred, you might ask the question: Well, maybe not? But the reality is, we've got fingerprints all over the crime scene," said Chadderton. "They're on the body. They're on the knife. . . . They're smeared all over the place and they're on the handle of the door into the house. It's just like, come on people . . . the reality is the body of the evidence says we're dealing with live fish!"

Yet for now, the electric barrier is the only defense between the Asian carp–infested Mississippi basin and the Lake Michigan shoreline, and this should make no one feel comfortable. In early 2014, I obtained from the Fish and Wildlife Service a video taken by an underwater camera at the barrier several months earlier. Just one three-minute clip revealed dozens of little fish swimming upstream through the swath of electrified water. Lab tests conducted on a scale model said this was not possible with the voltage the barrier is operating at, but nobody before had ever bothered to drop a sonar-like camera in the water to see what might actually be in the canal.

The Army Corps brass remain nonplussed.

"Those aren't carp," Peabody's replacement, Brigadier General Margaret Burcham, told me in early 2014 in a conference room overlooking the downtown Chicago River—the very stretch of river that the migrating carp would pass through on their way into Lake Michigan, a mile upstream. Burcham isn't a fish expert; she has a graduate degree in computer science. Actual fishery experts weren't nearly as confident in what Burcham claimed to see—or, more specifically, claimed to not see.

"You can identify that they are fish. You can identify that they're moving—you can see that," Aaron Woldt, Fish and Wildlife's deputy assistant regional director for fisheries, told me. "But you can't tell [which] species."

The day an Army Corps general becomes convinced that an Asian carp invasion of the Great Lakes is under way—the day the fish regularly start turning up in nets above the electric barrier—is also a day it will probably be too late to do anything about it.

A tiny private fish pond in Missouri offers a distressing glimpse of what might be in store for the Great Lakes. The owner had stocked his personal fishing hole with catfish, bass and bluegills. He was pumping it full of fish food, yet the fish appeared to be starving. So in early 2010 he called in a pond consultant.

"They came out with electrofishing gear, caught some fish and looked at them," the Geological Survey's Chapman said. "The fish were emaciated and he didn't know why. He said, 'There's something wrong here. We need to start over again.' They brought in rotenone and completely killed the pond."

Over the next week, the rotting carcasses of about 300 bighead carp surfaced. The smallest were 20 pounds. The big ones were a border collie–sized 35 pounds. "They looked like submarines coming out of the water," said Chapman, who arrived on the scene in time to watch the last of the dead specimens surfacing. "They'd breach!"

Poisoned Asian carp, Chapman explained, are different from many fish species in that they typically don't float to the surface unless the water is warm enough for gases to build up in their bellies, a process that can take a week. Or it may never happen if the poisoning occurs in cold weather, as was the case during the December 2009 Chicago canal poisoning.

It turned out, Chapman explained, that a decade earlier the previous property owner had stocked the tiny pond with a colony of bighead carp, one that eventually flourished right under the nose of the new owner, who had smelled trouble—but couldn't see a thing.

Chapter 6

CONQUERING A CONTINENT

THE MUSSEL INFESTATION
OF THE WEST

The St. Lawrence Seaway invasive species problem is not just contained to the shores of the Great Lakes. The problem is as big as North America itself. "The Great Lakes are just a beachhead for invasions that are going to play out across the country in the next century," says University of Wisconsin ecologist Jake Vander Zanden. "It's just the start."

And the biggest reason why is the Great Lakes' back door: the Chicago Sanitary and Ship Canal.

The story of the inland spread of zebra and quagga mussels perfectly illustrates the problem. When the first zebra mussel turned up in North America in 1988, biologists knew the fast-reproducing mollusks native to the Caspian and Black Sea basins would spread, they just didn't know how fast—or how far. One year after a group of Canadian college students doing field research stumbled upon the first invasive mussel specimen in Lake St. Clair, the small lake in the river system connecting Lakes Huron and Erie, the foreign mollusks that hitched a ride into the lakes aboard a freighter sailing up the St. Lawrence Seaway were found smothering the bottom of southern Lake Michigan near Chicago at densities exceeding 4,000 per square meter.

This was an expansion in range almost as stunning as the mussels' jump across the Atlantic Ocean. Freshly hatched mussels, called veligers, do have a brief, highly mobile stage when the microscopic juveniles coast on currents, like pollen in the wind. But a baby mussel can't paddle in the water any more than a pebble can do the backstroke, and the current coursing through Lake St. Clair would have pushed any drifting veligers downstream toward Lake Erie. The only plausible explanation for their 600-mile leap in the other direction was that the mussels had, once again, hitched a ride on a freighter.

Perhaps these Chicago mussels were unrelated to the initial St. Clair population and had been delivered straight to the southern tip of Lake Michigan from Europe by a different oceangoing vessel sailing up the St. Lawrence Seaway. Or perhaps they were transported west by a 1,000-foot-long "laker," a class of ship too big to squeeze through the locks of the Welland Canal and sail below Niagara Falls. Although these Great Lakes–bound freighters can't be blamed for introducing exotic species from foreign ports, they have a ballast water capacity about double that of a Seaway freighter. A single one of these ships could have easily sucked up millions of veligers in the freshly infested waters of Lake St. Clair and then seeded them in Chicago. The trip would have taken less than two days.

Had the Chicago mussels found the same Lake Michigan that explorers Jacques Marquette and Louis Joliet did in 1673, they would have hit a dead end. At that time, even water at the far southern end of Lake Michigan flowed slowly north, eventually east, through Lakes Huron, Erie and Ontario and out to the Atlantic Ocean. But the mussels that turned up in Chicago in 1989, of course, didn't find a dead end at the southern end of Lake Michigan; they found the Chicago Sanitary and Ship Canal that links the lake to the Mississippi River basin, which sprawls across 40 percent of the continental United States.

In spring of 1989 a biologist for the Chicago sewerage district doing a routine survey in the canal just southwest of downtown scooped three or four pea-sized shells from the sediment. He sent them off to the same Canadian mussel expert who had identified the first zebra mussels found in Lake St. Clair the year before, and he got the same answer. What ensued in the next few years was a veliger blizzard down the canal and into Mississippi River tributaries that nobody could have predicted. Biologists in the early 1990s calculated that the microscopic mussel veligers were tumbling down the Mississippi-bound Illinois River at a rate of 70 million per second. Closely related invasive quagga mussels were not far behind.

Still, if river currents were the only means for these mussels to spread inland, then gravity would have dictated that they stay confined to the course that Lake Michigan water takes on its trip toward the Gulf of Mexico—down the sewage canal, into the Des Plaines River, then into the Illinois River, and finally into the Mississippi River where the Illinois River joins it just outside of St. Louis. The Mississippi River upstream from St. Louis, along with all other Mississippi tributaries, would have been spared the invasion because mussels can't swim upstream. But it turned out zebra and quagga mussels don't need currents or even swimming pool-sized ballast tanks to hitch rides. The natural superglue they excrete to bind to hard surfaces allowed them to attach to the hulls of inland barges and other boats and ride them like elevators up and down the Mississippi River and all the navigable rivers and canals that connect to it.

By 1992 the invading mussels were found 800 miles upstream from where the Ohio River flows into the Mississippi, and they weren't just spreading in the Mississippi River basin. By 1993 they were found in Vermont's Lake Champlain, which is connected via canals to the

Great Lakes and to the Hudson River, which researchers estimated at that time was already home to more than 500 billion invasive mussels. By 1994 the invading mollusks had turned up as far south as Louisiana, as far west as Oklahoma and as far north as Minneapolis.

They were also soon popping up in dozens of isolated inland lakes with no direct connection to the Mississippi River or the Great Lakes. How could this be?

CANDY DAILEY WAS SPLASHING ON THE BEACH OF LAKE METONGA, a small northern Wisconsin lake, with her grandchildren over a Fourth of July weekend several years back when she felt a wicked sting. She didn't have to look down to know her foot had been sliced by a zebra mussel shell. "I'm a nurse, so I knew to make it bleed and wash it out," Dailey told me. "I dried it off and taped it."

The paper cut–thin lacerations, typically invisible until the first bead of blood arrives, have become as common as bee stings for swimmers on Dailey's lake since the invading mussels mysteriously made the 100-mile jump inland from Lake Michigan in July 2001.

Dailey didn't think much more about the mishap until the middle of the night when she awoke with a throbbing, swollen foot. By morning, a telltale red streak was creeping up her leg. By the following night, she was popping broad-spectrum antibiotics. Dailey recovered from the infection but her holiday was over. This, of course, is not the kind of story that merits even a brief mention in the local newspaper. It was just one common bacterial infection from one cut suffered by one person swimming on one lake. The problem is that Wisconsin alone has more than 15,000 inland lakes—and every one of them has a stake in stopping the next contaminated vessel from sailing up the Seaway.

Standing in front of a group of some 400 inland lake property owners in downtown Green Bay, Wisconsin several years ago, University of

Notre Dame biologist and invasive species expert David Lodge dimmed the lights and gave a short, chilling presentation about how biologically contaminated vessels sailing up the St. Lawrence Seaway threaten even the most isolated inland lakes.

"If you want to know what's coming next," Lodge told the gathering, "look at the species that are already in the Great Lakes."

He showed a slide revealing that the Great Lakes are directly connected to 12 percent of the world's ports via existing shipping routes. That means a mussel, fish or even virus picked up in, say, Antwerp, Belgium, could arrive in a matter of days at the Green Bay docks just outside the doors of the conference center at which Lodge spoke. But the murmurs in the crowd didn't start until Lodge pulled up his second slide, which revealed 99 percent of the world's ports are within two stops from the Port of Green Bay, or any other commercial dock in the Great Lakes. This, Lodge, explained, is more than a Great Lakes problem, because rarely do Seaway invaders stay contained to the big lakes. Invasive fish can swim up the tributaries feeding the lakes. Foreign microbes can infect baitfish populations that are then distributed to inland tackle and bait shops. And then there are the quagga and zebra mussels.

More troublesome than the mussels' reproductive rates and ability to bind to boat hulls is their virus-like ability to spread unseen in cagey ways biologists were not prepared for. It turned out the Caspian invaders have a remarkable ability to survive out of water. Depending on air temperature and humidity, a specimen can live for days on a boat pulled from the water. They may live longer by lurking invisibly in bait buckets, in the honeycombed foam of a wet sponge or even on the back side of a damp lifejacket strap. Put an infested boat on a trailer, and an otherwise immobile mussel can roll at blitzkrieg speeds. This is almost surely how the mussels colonized Dailey's Lake Metonga, and why they are now reported in waters of 28 states. In fact, the infestation continues to move so swiftly that the federal government now posts a daily update of affected lakes.

Yet for the mussels' first two decades in North America there was one geographic obstacle that they could not breach: the Rocky Mountains.

IT WAS A GUSTY, UNSEASONABLY COLD—LITERALLY FREEZING— day on the Nevada high desert in January 2007 when a team of maintenance men in scuba gear dropped to the bottom of Lake Mead for a pre–boat season inspection of cables and anchor blocks at the floating village that is the Las Vegas Boat Harbor & Lake Mead Marina.

Lake Mead is the nation's largest reservoir and holds enough water to submerge an area the size of Connecticut under 10 feet of water. The lake didn't exist until the 1930s, when an army of some 5,000 workers under direction of the U.S. Bureau of Reclamation finished construction of Hoover Dam, the centerpiece of the mammoth plumbing project that is today's Colorado River. In its natural state, the Colorado's waters started as melting snowflakes in the mountains of Wyoming, Utah and Colorado on the western side of the continental divide and, once gathered in the main river channel, coursed through the West's red rock country before plunging into the Grand Canyon on its rush toward Mexico's Gulf of California, taking with it all the raindrops and melting snow that falls upon a region the size of France.

Although the idea behind the St. Lawrence Seaway was to carve a new seacoast from the heart of the continent, the federal government's designs for the Colorado River in the early 20th century were even more audacious. The plan wasn't merely to deepen and widen a waterway to open an inland region to navigation. It was to consume a river, to use it first to spin hydroelectric turbines to supply electricity to millions of people and then to suck the river dry—every single drop. Today there are more than 53 dams on the Colorado River and its tributaries and the result is that, in most years now, the river disappears into the desert before it reaches the ocean. It bleeds into tunnels, aqueducts, ditches and pipes that carry it to the faucets of some 25 million people. The tamed river turned dusty desert outposts into the cities of Phoenix, Las Vegas, Tucson, Albuquerque, Salt Lake City, San Diego and Los Angeles. But even now most of the water taken from the Colorado River mists and drips from irrigation spigots that give life to some two million agricultural acres previously too dry to support much more than sagebrush.

The Southwest as we now know it would not exist had people not figured out how to move Colorado River water through valves, screens, grates and gates to put it predictably and precisely where they want it, when they want it. The mussels now threaten to gum up this entire grand plumbing system, and it all started with a couple of odd shells that looked as innocuous as sunflower seeds.

"We didn't think much about it," marina general manager Bob Gripentog said. "There was just one or two. Literally." But Gripentog knew enough to turn those specimens over to National Park Service staff who had been fretting that the mussel scourge of the Great Lakes would someday come tumbling down the western side of the continental divide. They sent the specimens overnight to a mollusk expert for identification and they got the answer they didn't want just two days later.

Nobody knows exactly who launched the quagga mussel infestation of Lake Mead, but it's a sure bet the mussels arrived attached to the hull or lurking in the bilge water of a Midwestern pleasure boat pulled across the Great Plains, over the Rockies and down one of the lake's dozens of boat ramps. These particular pioneering mussels could not have landed in a worse place. Lake Mead receives more than seven million visitors annually, almost as many as Yellowstone and Yosemite

National Parks combined. But unlike those other two national destinations, boating is the primary draw at Lake Mead and on busy weekends it can be buzzing with as many as 5,000 pleasure craft, whose owners come from states all across the West. If you were to draw a chart showing how biological invasions spread that looked like a wheel, Lake Mead would be the hub. The boats would be thousands upon thousands of spokes.

What took years to unfold in the Great Lakes started to play out in a matter of months at Lake Mead, in whose warm waters quagga mussels can reproduce six or seven times a year, compared with once or twice per year in the Great Lakes.

Barely 24 months after the divers' discovery, Lake Mead's once-brown canyon walls beneath the water line had turned black as charcoal with mussel shells. Divers found them smothering everything on the lake bottom, from beer cans to a downed B-29 bomber that crashed after World War II while taking atmospheric readings for a classified plan to develop a sun-guided intercontinental ballistic missile guidance system. The pilot's job on July 21, 1948, was to dive that bomber from 30,000 feet to just above the surface of Lake Mead. The lake's glassy reflection reportedly fuddled the pilot's depth perception and the belly of his plane hit the water at about 230 miles per hour, skipped back into the air and then crash landed again, this time softly enough for the crew to escape before the plane drifted to the reservoir bottom nearly 200 feet below, where it remained lost to the world until it was discovered by divers in 2001. Eight years later it had disappeared again under a blanket of mussels.

But the mussels weren't just spreading across the manmade lakebed and its canyon walls. The early days of the invasion pushed U.S. Bureau of Reclamation researcher Leonard Willett into a window-less office at the base of Hoover Dam to try to figure out how to beat back the plaque-like mollusks clustering in the dam's waterworks with chemicals, heat or even bacteria. Mussels smothering the face of the 726-foot-high concrete dam weren't the big problem. It was their ability to gum up the cooling system for the dam's whirring turbines built to supply electricity to some 1.5 million people.

"You wouldn't think that a little fingernail-sized mollusk could stop or slow down a dam this size, but when you see what these little critters can do, it is amazing," Willett yelled through my earplugs during a tour through the bowels of the dam. "They can quickly start shutting down even the largest infrastructures."

Later that day, the director of the Southern Nevada Water System pulled up a video on his computer screen that had recently been taken by scuba divers that showed mussels smothering the grates over two giant pipes that feed Lake Mead water to some two million people in the Las Vegas valley below. There wasn't a single shell on those grates the previous winter. Less than a year later there were millions if not billions of them and Las Vegas' public water supply would have been lost if divers weren't dispatched to scrub the shells away.

As on the Chicago sewage canal, the Western mussels quickly rode river currents to expand their range into downstream Colorado River reservoirs, dams and pumping facilities in Arizona and California. Within two years, the quagga mussels had colonized nearly three dozen new water bodies across Nevada, Arizona, California, Colorado and Utah. By early 2014, the bottom of Davis Dam straddling the Arizona–Nevada border downstream from Hoover Dam had been turned into an oversized laboratory to develop new medicines for what some regional Western leaders had begun to call "the STD of the Sea."

The Bureau of Reclamation's Willett had been reassigned to the Davis Dam so he could have more room to conduct his anti-mussel experiments, which included copper ion treatment systems designed to kill mussels with doses of metal measured in parts per billion. He tested a product that uses dead cells from microbes that the mussels eagerly gobble up as food. The cells then kill by destroying the mussels' digestive tracts. He experimented with UV light systems similar to those used to purify drinking water, and with high-tech filters and with paints that make it difficult for mussels to bond to surfaces. More than seven years after the quaggas had made the jump across the continental divide, Willett said he was confident a combination of these tools will allow Westerners to keep water flowing, though it is going to cost a staggering amount of money over the coming decades.

"There is plenty of opportunity to control them in a pipe," he said. "But I don't have any answer for how to eradicate them. I certainly can't control them in open water."

Putting an exact price tag on what this all will ultimately cost the West is impossible, but it's already well on the way into the hundreds of millions of dollars. Just seven years after the first quagga turned up in Lake Mead the Metropolitan Water District of Southern California alone reported it had already spent nearly $50 million to cope with the infestation. Its managers even pulled the $2 billion Diamond Valley

A sign on the shore of Lake Mead.

Lake Reservoir, completed in 1999 to provide a backup water supply to Southern California, off the Colorado River canal system. Water managers are now filling the reservoir with water piped down from quagga-free sources in northern California, but if that reservoir succumbs to an infestation, managers expect it will cost ratepayers about $2 million per year to keep water flowing.

The mussels have yet to make the jump into the Pacific Northwest's Columbia River basin, which drains an area larger than the Colorado River watershed. But the economic threat mussels pose to the Northwest is even greater because of the vast hydroelectric dam network along the Columbia and Snake Rivers and the billions of dollars the region has invested in trying to restore its beleaguered salmon stocks with hydroelectric dam modifications, habitat restoration projects and hatchery programs. Pacific Northwest politicians and their Western Canadian counterparts figure an invasion could cost their region collectively up to $500 million—annually.

This is not a surprise to the residents of the Great Lakes states, where the toll of the invasion has already climbed into the billions of dollars. Yet in many ways Midwesterners have learned to live with the scourge. They've grown accustomed to higher utility bills and to wearing shoes while swimming. They've become numb to buying exotic farm-raised tilapia and salmon instead of local lake fish at grocery stores and restaurants. They plug their noses, shrug and walk away from beaches fouled by rotting seaweed slicks triggered by the mussels. Most never think about the cause of it all, and if they do it is often written off as just the price that had to be paid for opening the Great Lakes to the rest of the globe.

With similar threats now hanging over them, Westerners aren't willing to consider someone who transports a mussel into new waters merely happenstance or an unfortunate cost of doing business.

They consider it a felony.

≋

AT THE PAGE, ARIZONA, AIRPORT NEAR SOUTHERN UTAH'S LAKE
Powell is a giant yellow sign that dwarfs all the usual security cautions
about firearms, flammable liquids and toothpaste gels. "DON'T MOVE
A MUSSEL," screams the poster just outside the terminal. "NOW IT'S
THE LAW."

The message is repeated on billboards all across the West as state
legislatures get tougher on carriers of biological pollution in a manner
that, were it applied in the Great Lakes region, just might push overseas
shipping on the Seaway out of business. States like Wyoming, Utah,
Arizona and California all have laws making it illegal to possess zebra
or quagga mussels, and law enforcement officials have little sympathy
for anyone caught accidentally transporting them across state lines,
something considered a criminal activity.

Nowhere has the fight to stop the mussels been more ferocious
than at Lake Powell, which, despite its remoteness, draws some three
million visitors annually. Lake Mead is a bigger reservoir and pulls in
more visitors than Lake Powell, but that is largely because of its proxim-
ity to Las Vegas. Lake Powell is nearly a five-hour drive from Las Vegas
and more than four hours from Phoenix, yet it is one of the nation's
most popular boating destinations because of its otherworldly, twisting
red rock canyons and fish-filled waters.

Biologists knew even before the mussels invaded Lake Mead, 370
miles downstream, that Lake Powell's national popularity with boaters and anglers made it particularly vulnerable to an invasion. In 1999,
National Park Service workers surveyed the boat trailers in one of the
lake's marina parking lots and counted more than 70 license plates
from states known to be mussel-infested.

"That suggested a pretty significant risk," Mark Anderson, a
National Park Service aquatic ecologist at Lake Powell, told me. The

next year, Park Service rangers started asking visitors if their boats had been in a mussel-infested state within the last 30 days. If so, the boats were further inspected and, when necessary, decontaminated with a pressure wash at a mussel-cooking 140 degrees. Yet the rangers worried some boats might be slipping through the voluntary screening program. Those fears turned to panic in 2002 while Anderson was giving a presentation to marina workers about the dangers posed by boats from places like Ohio, Michigan and Wisconsin.

"Hey," one of the marina workers piped up. "We just launched a boat from Wisconsin." Anderson stopped his lecture and scrambled with the class down to the dock. He turned the moment into a hands-on clinic on how to frisk a boat hull for mussels, during which he felt the telltale bumps beneath the waterline.

He got on the phone and was soon in touch with the owner of the big blue and white Chris-Craft cruiser dubbed *The Virginian*—home port Racine, Wisconsin, about 30 miles down the Lake Michigan shoreline from Milwaukee. It turned out the boat, recently purchased by a Colorado resident, had been out of the water for nine months prior to its Lake Powell launch. That lag meant the sunflower seed–sized shells affixed to *The Virginian*'s bottom were long dead before the boat arrived at Lake Powell, but Park Service officials nevertheless took the close call seriously.

The next year Lake Powell's voluntary boat inspection program became mandatory. Checkpoints were set up near boat ramps and boats deemed high-risk were required to get decontaminated with a blast of hot water. Even as nearby reservoirs in Arizona, Nevada and California fell to the mussel infestation, water samples regularly taken at Lake Powell continued to come up clean of DNA traces and microscopic mussel offspring. This was probably more than a matter of luck; in 2010 Park Service workers inspected 15,000 vessels and ordered 5,000 of them decontaminated. They also found 14 boats infested with mussels—before they were able to launch in Lake Powell.

Lake Powell, a man-made reservoir and popular boating destination that's now at the front line of the fight against "the STD of the Sea"—invasive mussels.

By 2012 those numbers had jumped to over 20,000 inspections and 6,000 vessel decontaminations, something that, depending on the size of the boat, can take more than six hours. That was the same year Wisconsin boat hauler Dwight Regelman rolled through the Lake Powell town of Page, Arizona, and pulled over just inside the Utah state line. Regelman had been hired to haul a 70-foot-long dinner cruise boat called the *Fiesta Queen* from Laughlin, Nevada, below the Hoover Dam to Saskatoon, Saskatchewan and was waiting for his Utah state trooper escort and other Utah transportation permits to clear when a swarm of wildlife officers descended. They started to climb all over the *Fiesta Queen* and they found what they were looking for—dozens of quagga mussels. They asked Regelman whether he knew he was illegally hauling quagga mussels into Utah. Regelman said he never heard of quagga mussels, but that he had the boat decontaminated for zebra mussels when it was pulled from the water in Laughlin. An officer invited Regelman to step into his vehicle for "a visit," then told him that he first needed to read to him something from a card.

"The reason I need to do this is because you haven't been free to leave, so I need to read to you the Miranda warning, which I'm sure you've heard a thousand times," the officer said in an aw-shucks tone that comes through crystal clear on the audio recording he made that day.

Regelman told me in a later interview that he thought he had done everything the law required before he trucked out from Wisconsin. He had called all the states he would be passing through to learn about their mussel regulations and, although he found the patchwork of laws confusing, said he was left with one take-away: the *Fiesta Queen*, which had been docked for years in mussel-infested waters, had to be thoroughly scrubbed before it rolled down the highways.

He paid $125 to a Bullhead City, Arizona, firm specializing in cleaning cars, trucks and RVs to give the *Fiesta Queen* a good scrubbing. Then he headed for Canada. "We were told it had to be pressure-washed and cleaned up and that's what we did," Regelman said. "We took the best information we could get and went."

The Utah wildlife officials who conducted the roadside interrogation weren't impressed. "Did you see mussels on it when you guys pulled it out?" an officer asked Regelman.

"Oh yeah, there was some on it," Regelman replied.

"And then they decontaminated it," the officer said. "Did you see any after that?"

"Umm," Regelman replied. "We went around and picked and dug some off, because yeah, there was some left after that. Not many but . . . still a few."

Utah law prohibits the transport of quagga mussels—dead or alive—through the state. Violators are subject to thousands of dollars in fines and jail time. The officer asked Regelman to take a walk around the *Fiesta Queen* with him. There were pockets of mussels all over, something that did not surprise Regelman, especially for such a large boat that uses river water for everything from its toilet flushes to engine cooling to air conditioning.

"If you're going to make sure there's none in there, you're basically going to have to scrap the boat or completely disassemble it," Regelman told the officer.

The officer did not disagree. It was decided the best thing to do was to let Regelman go back to Wisconsin while Utah wildlife officials put the *Fiesta Queen* in boat jail—a 30-day high-desert quarantine. About a month later, Regelman returned to Utah and proceeded north and delivered the decontaminated boat to its new owner in Saskatchewan, where it was re-christened the *Prairie Rose*. He returned to Wisconsin and said he did not give the episode much thought until the next year when Utah lawyers trying to scrounge up business began to call asking him if he needed legal representation; it turned out there had been a warrant issued for his arrest.

"We were charged with transporting illegal species. It carried a $5,000 fine," he said, "and a year in prison."

He was offered a deal to plead guilty and pay $5,557.34 in restitution for decontaminating and quarantining the boat. The settlement called for no jail time. Regelman did not find this reasonable.

"He really minimized the issue," Kent Burggraaf, who prosecuted the case for Kane County, Utah, told me as he leafed through the court filings in his cramped office in the desert town of Kanab, Utah, about 50 miles of sagebrush west of Lake Powell. "He didn't think it was a big deal. Well, it's a big enough deal that the Utah legislature passed a law."

Regelman finally agreed to plead guilty and pay $4,500 in restitution. The case was settled in early 2014, and Burggraaf said the guilty plea will be erased if Regelman stays out of trouble in Utah for a year. Regelman said the actual cost to him in terms of lost time, paying for staff and equipment and all the travel time added up to about $30,000.

He said he can appreciate laws to stop invasive species from spreading, but said they need to be reasonable.

What's unreasonable, Burggraaf and other Western law enforcement officials argue, is tolerance for even a single new water body to become infested. Burggraaf said he would similarly prosecute anyone else caught bringing quagga mussels into his state—dead or alive. "The law is still on the books, and until it's taken off, people need to decontaminate," he told me. "We're trying to keep it from spreading."

As he was telling me this it dawned on me that the day before a fisheries biologist had given me two dead mussel specimens as a souvenir of my trip out West. I put them in my reading glasses case the night before and didn't think any more about it until that moment in Burggraaf's office.

I had little doubt that I was in danger of getting arrested myself had I revealed to him the contents of the backpack that was sitting next to his desk.

WITH THE FIERCE FIGHT UNDER WAY TO KEEP MUSSELS FROM further metastasizing across the West, it is Lake Mead in Nevada—not the Great Lakes—that draws the most ire from Western politicians.

Some want the National Park Service operators of Lake Mead—Patient Zero in the region's spread of the STD of the Sea—to slow the scourge by requiring decontamination of every boat that leaves the lake. Yet on a busy weekend morning there can be three-hour waits at all the lake boat launches across its 550 miles of shoreline. Imagine, biologist Ryan Moore said, if each of the thousands of boats on their way back up those ramps were required to get decontaminated with a superheated pressure wash, the lines to leave the lake could last for weeks. And still, this strategy would almost certainly fail.

Moore recalled how one of Lake Mead's volunteer mussel experts taught a class on how to give a boat a proper hot wash. The volunteers spent hours on it, and then left the boat overnight, on dry land. The next day a couple of mussels had apparently used their single "foot" to drag themselves out of a crack or a cranny and were there in plain daylight, on the boat hull that had been declared mussel-free the day before.

"You can't remove every single mussel," Moore said, shaking his head at the absurdity of it, as if he were being asked to recite the names of each Lake Mead visitor the previous year. "You just can't."

Others have suggested quarantining boats for a month before allowing them to leave, including the mammoth house boats common on the lake. "It's not like you can just say 'we're going to impound this boat. Come back in 30 days,'" said Moore. "You'd have to itemize everything down to the sheets, knives, cups and forks, because there is always a risk you're going to return the boat and somebody will say, 'Hey my stereo is missing.'"

Moore said others pushed the Park Service to drain the lake, something Moore said his agency has no authority to do, let alone the fact that it could leave millions of people without drinking water. That scheme also ignores the fact that the mussels would still be clustered on the downstream side of the dam—and all the reservoirs below—and could recolonize the lake with the launching of just one contaminated vessel. There was also a suggestion that volunteer divers be deployed to hand-pick them out of the lake.

But maybe the wildest idea came from the boss at the Bureau of Reclamation that runs Hoover Dam. In early 2009, Lorri Gray, then regional director for the bureau, was fretting for the farmers downstream from the dam whose irrigation water was threatened. "The last thing they need," she said, "is another operation and maintenance cost or burden."

That had her thinking big. She was thinking it might be a good idea to bring in some sterilized mussel-eating black carp, one of the four species of invasive Asian carp now loose in the Mississippi River basin that are a threat to invade the Great Lakes by swimming up the Chicago Sanitary and Ship Canal. "That may be a good thing or a bad thing," she told me. "But that's something we're willing to explore."

The idea went nowhere, but the same can't be said about the mussels.

Officers trying to protect upstream Lake Powell over the next several years wrote more than 1,200 citations. Fines of $2,500 were common, even for local residents, people whom everyone knew boated only on Lake Powell. Lake Powell biologist Mark Anderson said every boat got treated the same, whether it was a $1 million floating mansion or a personal watercraft. He remembers sitting at home listening to his Park Service radio not long after the mandatory inspections started and hearing an employee say she was sending a family over to the decontamination station because they had a couple of small watercraft that needed to be washed. Real small. "They sent over swim noodles," Anderson told me. He had the proud look of a platoon leader talking about how one of his young soldiers had performed in battle. "I thought: 'Yes. Now they're getting it.'"

In 2012, park staff had already quarantined 38 boats when evidence started to trickle in that somehow, somewhere along the line, even with all the inspections, and decontaminations, court cases and fines, at least one boater had slipped through and contaminated the lake.

Biologists had no mussel shells to confirm it, but water samples showed the presence of microscopic veligers. Genetic sampling that can sniff out traces of the mussels' DNA also came back positive. Yet these lab results, worrisome as they were, never led scientists to colonies of adult mussels. Then in March 2013 a local businessman working on a boat that had just been pulled from Lake Powell called the Park Service to report four suspicious shells clinging to the boat's hull. These calls had been coming in for years, and they always turned out to be something else—a different type of native mussel or a snail or a shard of some unidentifiable shell. "I expected it to be another one of those," Anderson said. "But, definitely, they were quaggas."

The Park Service called in 36 divers in a "mussel blitz" to scour the boat and dock bottoms in the area where the infested vessel had been kept. After four days, the divers returned with a scant 235 mussels, ranging from pinky nail- to nickel-sized. But the more Anderson's crew looked in the following months, the more they found. They counted every specimen until finally surrendering in spring 2014 after a single boat pulled from the water had more than a thousand mussels freckling its bottom.

"We're disappointed," Anderson told me just days afterward. "Distraught, really."

Anderson was at that moment just beginning to grasp what untold generations to come had suddenly been stuck with due to failed invasive species measures on the other side of the continent.

"If a 55-gallon barrel of oil was spilled in the lake, people would think, how awful—but you have to remember, you have nothing but 55 gallons. That's it," Anderson said. "But if you take two mussels and get them together, it can be mussels everywhere. Just. Everywhere."

Wayne Gustaveson, a biologist with the Utah Division of Wildlife at Lake Powell, had also given up the fight to keep the mussels out by spring 2014. He had moved on to trying to figure out how far the mussels had spread throughout the lake and once he realized they liked to cluster in shadowed sandstone overhangs, he started finding them everywhere.

Gustaveson took me on a boat trip just upstream from Glen Canyon Dam that holds back Lake Powell and pointed to an outcropping that he said looked like prime habitat. Sure enough, dozens of invaders were clinging under a red rock overhang along a stretch of canyon wall that had been underwater the previous year.

They were dead, dry as dust, but Gustaveson figured there were hundreds of live ones, if not thousands, directly below the waterline along the canyon wall that stretches hundreds of feet down. His prognosis for the lake's fishery was not good; he figured he had maybe 15 years until the fishery risked a Great Lakes–style collapse. The lake's food chain, Gustaveson explained, is pretty simple—plankton are eaten by shad, which are then eaten by introduced striped bass, two million of which are caught on Lake Powell annually. Once the mussels consume enough plankton to lower shad numbers, the stripers are in trouble. Gustaveson was already thinking a new ecosystem may have to be engineered. He hopes to find a fish that will eat the mussels and, in turn, be a source of food for striped bass—or some other species attractive to anglers.

"I'll do the best I can to restore that food-chain link when the mussels overtake the lake," he said as he glumly picked mussels off the canyon wall as if they were burrs on a blanket—a blanket that stretched to the horizon.

AFTER LOSING THE FIGHT TO SAVE LAKE POWELL FROM A QUAGGA mussel infestation in 2014, the Park Service briefly considered requiring boats *leaving* the reservoir to be decontaminated to prevent it from becoming yet another launch zone for invasions of other lakes and reservoirs. But, like Lake Mead, this is an impossible task given the millions of visitors Lake Powell receives annually. Park Service workers instead require visitors to decontaminate their boats on their own after exiting the lake, an expensive, time-consuming process at least some are certain to skip when no one is looking.

With mussels now established in Nevada, Arizona, California and Utah, the front line has been pushed north. Washington state law enforcement officials in early 2014 were preparing felony cases against two people caught hauling mussel-infested boats across their state line. In Idaho, regulators stopped 43,778 boats along state highways in 2013, all to catch the 12 boats that were actually found to be carrying mussels. This means that more than 99.99 percent of trailered boats are not a problem for Idaho.

This is a number similar to what Seaway operators boast has been accomplished by requiring overseas freighters to flush their ballast tanks with mid-ocean saltwater before discharging ballast into the Great Lakes. The shipping industry considers the resulting 98 to 99 percent or greater reduction of live organisms per gallon of ballast water as a success; Idaho regulators see their own 99.99 percent number differently. They operate under the principle that even one boat slipping through is enough to lose the fight. That is why Idaho inspectors were back at the highway checkpoints the next spring and that is why, by 2015, they had stopped and inspected 260,000 boats. This is expensive business. Altogether, Western states have been spending some $20 million annually to try to stop the spread of aquatic invasive species.

Yet the battle to keep mussels from infesting the Columbia River basin is one destined to be lost. In any given year there are hundreds of thousands of boats rolling down tens of thousands of miles of roads heading for countless unguarded boat ramps. Trying to catch every infested boat is like trying to handpick swarms of flies from an infested mansion. The trick, of course, is to not let those flies through the door in the first place, which brings us back to the St. Lawrence Seaway.

Although the Great Lakes have more than 10,000 miles of shoreline—more than the United States' Pacific and Atlantic Coasts combined—the region is uniquely positioned to stop biological invasions because of one incredible geographical fact: every overseas freighter that sails up from the East Coast into the lakes must pass through a single pinch point: the first lock on the St. Lawrence Seaway. It is 80 feet wide. That's a little broader than some busy city streets and a little narrower than the distance between home plate and first base. There is a boat ramp at Lake Mead that is wider.

Put a screen on this door, and your next "fly" problem is solved— for the Great Lakes, and for the whole continent. Yet essentially every day during the Seaway's nine-month shipping season this door swings open to overseas freighters, and to god-knows-what they might be carrying in their ballast tanks.

Although every overseas vessel passing into the Seaway gets stopped, the difference between Western wildlife officials and Seaway inspectors is that the Seaway inspectors aren't actually looking for invasive species. They merely test a ship's ballast tanks for salinity to ensure captains followed the requirement to flush the tanks with mid-ocean saltwater to expel or kill foreign freshwater hitchhikers yet to colonize the lakes, dozens of which have been identified by federal biologists as having the potential to still do so.

Seaway managers and shipping industry advocates argue that these salinity tests, paired with new regulations that will eventually require all overseas ships entering U.S. waters to install ballast water disinfection systems, are protective enough to stop new invaders. Yet it likely will be the better part of a decade before treatment systems are required for all ships sailing into U.S. waters, due to the time and expense it takes to develop and install the equipment. So for the foreseeable future, a salinity check remains the last line of defense for the lakes—and for all of North America.

But other tools are available right now to determine whether a boatload of trouble might be moving through the Seaway locks. Specific genetic markers, for example, have been developed to identify a number of organisms that researchers say are prime candidates to invade the continent, and that means ballast tanks could be screened for these invaders.

There are problems at this point in using this "eDNA" sampling technique (the same technology the Notre Dame team used to track the spread of Asian carp) on ships entering the Seaway. Water samples can take weeks to process, and a positive result doesn't reveal whether the species that shed its genetic material is still alive. Still, used in a research context, it could give Seaway operators a better idea of the potential trouble lurking in ballast tanks.

But the Seaway's U.S. operators are not interested in looking for DNA on their regular inspections of arriving boats.

"I think eDNA is emerging. They're making strides in that, from what I understand," Deputy Seaway Administrator Craig Middlebrook said. "But to say we'd go out and implement it tomorrow? We're not there yet on that."

That the Seaway would not take advantage of a DNA water test makes no sense to boat hauler Regelman. "Heck, they DNA'd my mouth," he told me, explaining that part of his plea deal in Utah required a mouth swab for a criminal background check. "I'm sure you could go onto any one of those ships and find invasive species—any single boat."

Especially the 47 of the 371 overseas ships that entered the Seaway in 2013 that tests showed had ballast tanks or other water tanks that did not meet the required salinity standard. Those ships' captains had to promise not to discharge any water from those tanks while in the Great Lakes. The tanks were then sealed so inspectors could board the freighters on their trip out of the lakes to make sure the promise was kept.

But what exactly did it cost a ship operator who violates that promise? Seaway officials reported in 2014 that it has happened only a couple of times since the rules were put in place, and that the maximum penalty is a civil fine of $36,625, although in 2011 the sole illegal discharger was fined $3,000. In 2010, a captain sailing in from Mexico who broke the promise was fined the same amount. Seaway officials told me in 2014 that this is a typical fine for a first infraction.

Three thousand dollars.

When I told Lake Powell ecologist Mark Anderson this, just days after he and his staff had finally given up the 14-year, $7.5 million fight to prevent a quagga mussel invasion of Lake Powell, he gave me a two-second blink. He could not grasp how ships capable of discharging millions of gallons of potentially contaminated water could be fined less than a local jet skier who skirted a checkpoint at the local lake. He had that look of a platoon leader once again—one who learns that the combat mission he just sent his troops on is senseless.

Three thousand dollars?

"And for that, we're risking things like these mussels getting everywhere?" he asked, his mouth agape.

"Wow," he said as he slowly shook his head. "Wow."

Two years later, water planners were already fretting that quagga mussels would gum up plans for a new 140-mile drinking water pipeline to fuel the population boom near southern Utah's St. George. And biologist Gustaveson was wondering whether his lake's striped bass population would survive the decade. It was the exact three-wave invasion sequence that transpired in the Great Lakes—discovery followed by water supply issues and then looming ecological chaos.

Chapter 7

NORTH AMERICA'S "DEAD" SEA
TOXIC ALGAE AND THE THREAT TO TOLEDO'S WATER SUPPLY

As there are lakes, and then there are the Great Lakes, there are swamps, and then there was the Great Black Swamp. While settlements west of the Appalachian Mountains exploded across the middle of the continent in the early 19th century, a morass on the western end of Lake Erie remained largely unsettled—if not impassable—for decades after Ohio gained statehood in 1803. Beneath the still water was a quicksand-like black mud that sat atop an impervious bed of clay. From the depths of this glacier-carved quagmire sprouted forests of ash, sycamores and elms tall and dense enough to block out the summer sun. Maples, hickories and oaks towered in the drier zones, and where the canopies of leaves gave way, there burst tangles of grasses so tightly woven they could stop an army in its tracks.

The Great Black Swamp was once called the "most forsaken, desolate and inhospitable" patch of wilderness in the nation. But patch is probably the wrong word. It was nearly the size of Connecticut—twice the size of Everglades National Park.

Ever-shifting in shape and size depending on the season, precipitation patterns and Lake Erie water levels, the swamp stretched west from the lake into eastern Indiana. On a map, it had the jagged borders of a worrisome mole. On the ground, things were more frightful. Ohio's swamp teemed with big cats and bears, wolves and wolverines, rats and poisonous water snakes. In the dank air that hung above the sheets of bathtub-flat water swarmed hordes of biting flies big as bumblebees and malaria-carrying mosquitoes that so routinely brought the fatal "shakes" to the few who settled on the swamp's edge in the first decades of the 19th century that the survivors comforted themselves with this grim ditty: *"There is a funeral every day, without hearse or pall; they tuck them in the ground with breeches, coat and all."*

The Great Black Swamp was such a geographical cipher that Native Americans at war with the United States in the late 1700s used it as a redoubt, confident the tender-footed soldiers chasing them had no interest or ability to navigate their way in, let alone out. It took General "Mad" Anthony Wayne (to whom Fort Wayne, Indiana owes its name) to slog his troops in and corner the surprised warriors. It was not much of a victory; the next year Wayne turned the swamp back over to his former enemy in the 1795 Treaty of Greenville that ended the decade-old Northwest Indian War.

The swamp was not a place where settlers could live in harmony with nature, so it was destined to be a place that they either left to nature, or a place that they took from nature (and the Native Americans they had ceded it to). It might well have been the former except that the swamp sat in the middle of the migration route for emigrants streaming from the East toward the fertile soils and lush forests of Michigan. The first road in, built in the 1820s, was soon abandoned as a useless muddy mess. A later road, built upon a latticework of logs, was still so sloppy that the overland trip between Cleveland and Detroit could only reliably be made in winter when it was frozen. And even then swamp travelers had no guarantee they'd make it out the other side. This is one man's 1838 account of his trip across the muddy abyss:

> We had scarcely got started when the wagon, getting off the beaten track, broke through the ice and mud, our fore-wheel going entirely below the surface of the road. This was a caution, and after prying out of this we avoided other holes, escaping accident ourselves, though we were compelled to witness trouble and its emblems every step of the way. Every few rods someone may be seen prying out a piece of wreck of some wagon or other vehicle. Many horses have been killed, and some men seriously injured on this road. When the frost is out, particularly in the spring and fall, it is one entire mud-hole . . . for 30 miles.

To establish even this ephemeral passage through the swamp required extensive ditch digging on each side of the roadway to channel away the standing water, and it didn't take long for settlers in com-

The Great Black Swamp, once a treacherous—and ecologically crucial—wilderness, was drained over the decades to create more arable land.

munities at the edge of the swamp to realize that the vast wasteland in their midst could be drained similarly with even bigger ditches that could tilt the driftless waters toward nearby rivers, including the Maumee River that flows through what is today downtown Toledo. Farmers sniffing cheap, fertile land soon flooded into the swamp with shovels and picks to construct an expansive network of ditches and subterranean clay pipes, called drain tiles, that would ultimately bleed the swamp dry as fields, fields bursting with corn.

The work began in the middle decades of the 1800s and the pace quickened in the 1880s with the locally invented Buckeye Traction Ditcher, a steam-driven contraption that chitty chitty bang-banged through the swamp, carving trenches for drain tile at such a pace that eventually 15,000 miles of manmade water channels would lace the old swamplands. "The people here treated the Black Swamp as though it were the enemy," observed one historian in the early 1980s. "And they annihilated it."

What was left at the dawn of the 20th century once the swamp was conquered was some of the richest soils on the continent because the cache of nutrients built up over thousands of years of life, death and decay had never been given a chance to drain away. Today the Maumee River basin is among the most intensely farmed regions on the globe. The watershed sprawls across a little more than four million acres, about three million of which is cropland. If you drive across these old swamplands today, where once sturgeon darted, snakes slithered and humming bugs hummed, you will encounter endless miles of zipper-straight rows of corn, now so ingrained in the economy, culture and landscape of the western Lake Erie basin that even many locals never bother to think about what it all once was.

The problem is that the ditch digging broke more than the figurative backs of the people who did it. In a way it also broke Lake Erie. In ecological terms, the muck and still water of the Great Black Swamp wasn't a wasteland at all. It was Lake Erie's kidney, a grand filtering system that turned muddy rainwater flushing off the land into crystalline flows by the time they reached the lake. And it was replaced with a drainage organ that does quite the opposite; the giant grid of pipes and ditches mainlines storm runoff—and all the wastes it contains—straight into the lake.

Lake Erie suffered immensely throughout the late 19th and 20th centuries as a receptacle for human, industrial and agricultural wastes. But nothing compares to what is happening today. Those millions of acres of destroyed wetlands, the overapplication of farm fertilizer, an increase in spring deluges and a lakebed smothered with invasive mussels have all conspired to create massive seasonal toxic algae blooms that are turning Erie's water into something that seems impossible for a sea of its size: poison.

IN 1831, NOT LONG AFTER THE FIRST DRAINAGE DITCHES WERE being gouged through the Great Black Swamp, French historian and writer Alexis de Tocqueville took a boat trip across Lake Erie, beyond the forests of the swamp, up the Detroit River and on into Lake Huron, where he was left grasping for words to convey the majesty of it all: ". . . this shore which does not yet show any trace of the passage of man, this eternal forest which borders it; all that, I assure you, is not grand in poetry only; it's the most extraordinary spectacle that I have seen in my life," the 26-year-old wrote to his father. "These regions, which yet form only an immense wilderness, will become one of the richest and most powerful countries in the world. . . . Nothing is missing except civilized man, and he is at the door."

He is lucky he made the trip when he did. Only three years later, Cleveland would open its first incorporated manufacturing plant, the Cuyahoga Steam Furnace Co., which would employ some 100 workers, and by 1850 the area would have a population of nearly 50,000. In the 1860s, less than four decades after Tocqueville first sailed across Lake Erie, Cleveland's Cuyahoga River had become so polluted it was already prone to burning, and by the middle of the 20th century Lake Erie was inspiring an entirely different kind of poetry.

It was Dr. Seuss himself who laid bare the national embarrassment that the smallest and most vulnerable of all the Great Lakes had become by the second half of the 20th century, lamenting in *The Lorax* a mythical place so polluted that "fish walk on their fins and get woefully weary in search of some water that isn't so smeary."

"I hear," Dr. Seuss wrote in his 1971 classic, "things are just as bad up in Lake Erie."

Seuss wasn't elegizing the fouled Cuyahoga River. Lake Erie had a far greater problem than digesting the human and industrial ordures oozing into Cleveland Harbor at the time. At more than 240 miles long and nearly 60 miles wide, the little Great Lake was still big enough that the Cuyahoga's toxic insults were largely a localized phenomenon; think of a gritty, chronically polluted urban beach in, say, the Bronx, and then think of the $15 million beachfront homes just a few miles across Long Island Sound. Or consider Indiana Dunes National Lakeshore, a 15-mile ribbon of glorious Lake Michigan swimming beach that draws two million visitors annually—and is only about 15 miles east of Gary, Indiana's industrially ravaged lakefront.

Seuss was talking about an environmental catastrophe on a grander scale, and it wasn't the biological pollution that had started to flow into the lakes from Seaway ships discharging contaminated ballast water; it would be almost two decades before invasive mussels would begin to take hold in the lake.

No, the smeariness Seuss bemoaned at that moment was that of a lake choking on an overdose of nutrients that are as fundamental to life as sunlight itself. In its pristine state, the waters of Lake Erie were already primed to sustain an abundance of algae—the foundation of its food web. The warmest, shallowest and furthest south of the Great Lakes, it was naturally loaded with the building blocks of life—among them carbon, oxygen, hydrogen, nitrogen, zinc, copper, calcium and silicon. This is why Lake Erie, which holds only 2 percent of the overall volume of Great Lakes water, is home to about 50 percent of Great Lakes fish.

But this biological richness left Lake Erie particularly vulnerable to an ailment called "cultural eutrophication," which is the condition of a pond or lake so overdosed with nutrients that have been put into the water by human activity (farming, sewage, lawn fertilizer, etc.) that the resulting explosion in algae suffocates other aquatic life. It is a condition that can, eventually, destroy a lake. As seasons, years and decades pile up, a body of water can be subsumed by all its accreting dead plant and animal life and eventually it turns into a bog, killed by all the things it gave life to. This pond-to-bog process happens naturally and is known as eutrophication; *cultural* eutrophication speeds the progression immeasurably, like throwing gas on a fire. On a lake the size of Erie, the immediate worry with cultural eutrophication is that the nutrient overdosing creates so much life that its inevitable decay burns up so much oxygen that almost nothing can survive.

Historically, what kept Erie's algae growth in balance with the larger life forms that consumed it—what kept the lake from becoming an over-vegetated cesspool—was the small but steady dose of phosphorus naturally trickling off the landscape. Phosphorus in Lake Erie and the other Great Lakes is what biologists term a "limiting factor" for plant growth; dial up the phosphorus load, and everything else is in place to boost algae production to such proportions that its oxygen-burning decay creates vast "dead zones" that can span hundreds and even thousands of square miles.

This is what was happening in the late 1960s, when people started to refer to Lake Erie not as a Great Lake, but as "North America's Dead Sea."

PHOSPHORUS DOES NOT EXIST NATURALLY IN THE ENVIRONMENT in its elemental form, but it is widely distributed in rocks that, over time, leach it into the living world that craves it like oxygen. Phosphorus is required by every living cell for things like energy production and storage, DNA replication and tissue growth. It cycles through organisms as simple as single-celled algae and as complex as the human body, which is a tangle of trillions of cells that, collectively, carry nearly two pounds of phosphorus, mostly in bones and teeth as well as the brain, liver, muscles and, to a lesser extent, kidneys, lungs and skin. But humans, through their diet, take in more phosphorus than they need and are therefore constantly shedding traces of it, primarily in urine. It was from this waste stream that element number 15 on the periodic table was first isolated by a Hamburg alchemist named Hennig Brand in 1669.

Brand made his discovery by boiling away buckets of urine until all that remained was a crusty residue. Then he kept the flames going under a piece of glassware that held the urine concentrate until the glassware filled with luminescent fumes. And from that vessel dripped a liquid that burst into flames when it hit the air.

The fiery liquid thick with a garlicky stench was captured in another piece of glassware. The flames subsided and the liquid congealed into a white, waxy blob, upon the surface of which cool, faintly green flames appeared to dance—for days.

Brand, a former military man with little or no formal medical or scientific training but with a bloated enough sense of self that he insisted on being addressed as Herr Doktor, called his "discovery" phosphorus, which roughly translated from Greek means "bringer of light."

He was convinced that what he had in his glassware was the philosopher's stone, a mythical substance alchemists believed could convert base metals into gold and silver. What he found, it turned out, couldn't make lead glitter. So his glowing nuggets were only a curiosity used to wow theatergoers and stoke the jealousies of fellow alchemists who could not figure out how their rival acquired such a magical substance.

Rival laboratorians eventually learned how to produce their own phosphorus and within a century the source material for the element became human and animal bones and teeth. Later, rocks rich in phosphates, which is a form of salt containing phosphorus, would be mined and processed for the mineral that doctors came to believe could cure everything from impotence (it couldn't) to tuberculosis (it couldn't) to depression (it couldn't) to alcoholism (it couldn't) to epilepsy (it couldn't) to cholera (it couldn't) to toothaches (it couldn't).

But as the centuries would unfold, it turned out phosphorus did make a hell of a lethal rat poison, a spectacularly combustible match tip, a dastardly battlefield gas and, in an unlikely twist of history and fate, a wicked class of incendiary bombs dropped by the Allies in World War II that killed tens of thousands of Germans in phosphorus' own hometown, Hamburg.

Tragically for Lake Erie, concoctions made out of the mineral had some peaceful applications as well.

Scientists reckon that around the year 1800, a quarter century before canal builders opened Lake Erie to a human population explosion, some 3,000 tons of phosphorus flushed annually into the lake. It seeped off the land from rocks, soils, and decaying plant and animal life. It also fell from the sky in raindrops that captured microscopic airborne phosphorus particles. By the 1930s the lake's annual phosphorus load had climbed to 10,000 tons. By the late 1960s some 24,000 tons of phosphorus flowed into the lake annually.

Some came from poor or nonexistent sewage treatment—an adult human excretes about one pound of phosphorus annually.

Some came from livestock wastes that farmers—many of them working the land that was once the Great Black Swamp—used to replenish the trace amounts of naturally occurring phosphorus in farm soils that had been stripped of the mineral by decades of crop harvesting. Later, those manure applications were augmented—and in many cases replaced altogether—by manufactured fertilizer produced from phosphorus-rich rock carried by trucks and rail cars rolling into the Lake Erie basin.

And some came from phosphorus-based laundry soap. Prior to World War II, the cleaning agent most commonly used in U.S. washbasins was simple bar soap that contained very little phosphorus. But in the late 1940s and early 1950s, as the war machine's assembly lines started to roll out household appliances, soap manufacturers eager to tap the burgeoning washing machine market began dosing their product with phosphorus. It had an uncanny ability to pull grease and grime out of clothing in a way that plain soap could not, and the new formulation was so successful that if you bought a box of laundry detergent in the 1960s, you were basically buying a box loaded with phosphorus. Tide was 12.6 percent phosphorus by weight; Bold 11.8 percent; Gain a full 10 percent and Cheer a smidgeon less. By 1967 scientists had calculated that wastewater laden with laundry detergent was responsible for some 12,500 tons of phosphorus flowing annually into Lake Erie, roughly half the lake's entire annual load and more than four times the amount that flowed off the landscape in Lake Erie's presettlement days.

This was a high biological price to pay to pull sweat stains out of work shirts. The excess phosphorus had so speeded up the eutrophication of Lake Erie that scientists in the early 1970s estimated that in the previous five decades the 10,000-year-old lake had "aged" an additional 15,000 years. They calculated that the amount of algae floating in the lake had increased six-fold between 1920 and the mid-1960s, a phenomenon that inspired the 1969 Peabody Award–winning NBC documentary: *Who Killed Lake Erie? Everybody did it. But everybody denies it.* Two years after that, Seuss had venom in his pen when he sat down to write what he would later describe as his favorite book. "*The Lorax* came out of my being angry," he said a decade later. "I was out to attack what I think are evil things and let the chips fall where they might."

The year after *The Lorax* was published the United States and Canada agreed to work together to throttle the amount of phosphorus pumping into the lake. Biologists calculated the annual load of phosphorus going into the lake from sewage treatment plants, industries and agriculture and then figured how much would have to be pulled from those waste streams to bring algae populations back to healthy levels. The scientists working on the problem were confident they could solve it because Lake Erie empties and refills itself—its so-called water retention time—about every two and a half years (most of its water flows in from Lakes Michigan and Huron, an input offset by the lake's outflow down the Niagara River and over the falls on its run to the sea). That means if they stopped the phosphorus overloads, the overdosed water should flush into the ocean in just a few years.

The agreement, signed in 1972 and updated later in the decade, called for annual phosphorous discharges into Lake Erie to be slashed by more than half, from an average of about 24,000 tons a year to 11,000 tons. Accomplishing this was not cheap. More than $8 billion was spent on sewage treatment plant upgrades. The states and province that shared ownership of the lake also passed laws slashing the levels of phosphorus allowed in soaps, and millions of dollars were spent working with farmers to keep phosphorus-rich fertilizer and manure from seeping off croplands.

The lake's recovery was as stunning and as science-driven as Vernon Applegate's lamprey control program a decade earlier. As predicted, the overall amount of algae in the lake shrank by about 50 percent as phosphorus loads dropped to target levels, and in less than two decades the lake's reputation rocketed from North America's Dead Sea to the Walleye Capital of the World.

The lake's turnaround was so dramatic, in fact, that in late 1985 two Ohio State University graduate students wrote Theodor S. Geisel—aka Dr. Seuss—and entreated him to remove his distressing assessment of Lake Erie from *The Lorax*, inviting Seuss to travel from his home in San Diego to Cleveland to see for himself the lake's revival.

In the summer of 2014, I contacted Rosanne Fortner, an Ohio State emeritus professor who specialized in environmental communication, to see if the university had any record of this letter. She told me that the invitation to Seuss had apparently been lost, but she did send me Seuss's 1986 response to the invitation, in which the doctor was contrite. The shot he took at Lake Erie in 1971, he acknowledged in a distinctly Seussian tenor, was no longer on the mark. "You must think me terribly rude for not answering your very pleasant letter of Dec. 6. The fault, however, is not mine. It just arrived this morning, having been somewhat circuitously forwarded from New York via pony express," Geisel typed on a piece of Seuss stationery.

"Although I will be unable to accept your kind invitation to come to Cleveland, I do agree with you that my 1971 statement in the *Lorax* about the condition of Lake Erie needs a bit of revision. I should no longer be saying bad things about a body of water that is now, due to great civic and scientific effort, the happy home of smiling fish.

"I can assure you the process of purifying my text will commence immediately. Unfortunately, the purification of texts, like that of lakes, cannot be accomplished overnight. The objectionable line will be removed from future editions. But it could possibly take more than a year before the existing stock of books has moved out of book stores."

≈≈≈

SEUSS, WHO MADE GOOD ON HIS PLEDGE, DIED IN 1991. WERE HE alive today, he'd probably be angry enough to put the line back. It turned out Lake Erie wasn't actually cured of its algal ills; they had only gone into remission. The lake's phosphorus load continues to this day to hit the target biologists set in the 1970s, but massive algae slicks mysteriously started to appear in the mid-1990s and the resulting dead zones eventually returned to levels rivaling the lake's dark days of the 1960s and early 1970s.

There are many species of algae in a freshwater body like Lake Erie and most are, at reasonable levels, essential to lake life. They convert sunlight into carbohydrates that are grazed upon by tiny animals like rotifers (which can be as small as one 1000th of a millimeter) to water fleas that can grow up to a half inch in length. That zooplankton is then gobbled up by small fish, which are eaten by bigger fish, and on it goes up the food web.

But there are also a variety of algae, commonly called blue-greens, that aren't really an algae but a primitive form of bacteria with the algae-like ability to soak nutrition from the sun through photosynthesis and can be packed with toxins, some of which pose a threat to humans. One particularly nasty form of blue-green algae that thrived in Lake Erie when phosphorus loads started to increase during the 20th century is called microcystis. It carries a poison so toxic that merely swimming on a beach where it lurks can induce vomiting, diarrhea and blistering around the mouth. If swallowed in sufficient quantities it can trigger liver failure. There have been no deaths attributed to it in the United States, but an outbreak in a public water supply at a dialysis center in Brazil in 1996 killed some 50 people. Its presence in Lake Erie waned dramatically following the phosphorus reductions of the 1970s and '80s but it has returned in recent years with a vengeance. One outbreak in

2011 was so big it took a satellite to track the poisonous blob, which from space looked like bright green psychedelic swirls set against the deep blue surface of the lake. At its peak, the algae slick spanned the western fifth of Lake Erie, smothering nearly 2,000 square miles of open water—more than three times larger than the previous record bloom. It was more than just a thin surface scum. In places the algae was nearly four inches deep and so mucousy it strained the motors of boats that tried to plow through it.

Two summers later, western Lake Erie's waters had turned so toxic with microcystis that it knocked out the water supply for a community of 2,000 residents east of Toledo. The event drew scant attention from the press but was recognized by public health officials at the time as an ominous moment for the Great Lakes. It was the first time a microcystis outbreak had overwhelmed a public water treatment plant.

It wouldn't be the last.

With Lake Erie's microcystis blooms in recent years having become as predictable and affixed to the rhythm of the seasons as dandelion blossoms and fall foliage, I decided in the summer of 2014 to go to Toledo to see for myself what had become of the Great Black Swamp, and what had become of the Lake Erie waters it once protected. On that five-day trip I stumbled upon the seeds of a natural and public health disaster unlike anything this country has experienced in modern times.

My first stop that Monday morning was at the lab of Tom Bridgeman, a University of Toledo freshwater biologist who is an expert on Lake Erie's resurgent algae troubles. Bridgeman has a sharp nose and intense blue eyes, which reminded me a bit of the character Hooper, the biologist Richard Dreyfuss played in *Jaws*. But the underwater menace Bridgeman was tracking posed a greater threat to the public than one hungry fish; great white sharks, after all, don't threaten children in their own bathrooms.

Bridgeman's office is on a piece of land carved out of the west end of Maumee Bay State Park east of Toledo, and his desk is not more than a couple hundred yards from the Lake Erie shoreline. We walked across the street to a beach smothered in a sludge so thick it killed the two-foot waves rolling in from the north before they could hit the shore. The rotting algal mat was not the potentially deadly microcystis. It was a stringy stew of blue-green algae called *Lyngbya* that has plagued Bridgeman's shoreline since 2006. It looks like rotten creamed spinach and, although scientists still don't know if it is an invasive species or a native strain that had been lying dormant for decades, they all agree it is yet another symptom of a lake that has been knocked dangerously far out of balance.

Bridgeman explained to me that the more dangerous microcystis wasn't visible from shore on this muggy, gray day. But he said colonies of it were surely lurking out in the lake's open waters, submerged, waiting for the wind to calm and the sun to come out for days on end to drive up water temperatures. Once these conditions set up, a probability by late August, underwater seed colonies of microcystis would float to the surface and explode into blooms covering hundreds—if not thousands—of square miles.

Bridgeman looked down at the beach and shook his head. His summer vacation was due to start the next week. He had spent his own childhood vacations swimming in Lake Erie, and he said he still relishes a dip in the lake when conditions are good. But there was no way he would have his family swim in this sludge. So, instead, he was headed with his wife into the rolling hills of central Ohio's Amish country to go camping.

We turned our backs to the lake and headed back to Bridgeman's laboratory, where he showed me a picture of a glass filled with a liquid so green it could have passed as a New Age health drink. It had been scooped from the lake a few miles offshore during a previous microcystis bloom, but this green concoction shouldn't be called water, not any more than a bottle of nail polish remover should be called water.

"You'd get really sick," Bridgeman said when I asked what would happen to someone foolish enough to take a gulp of the stuff. "I mean, drinking that whole glass might be enough to kill you."

He'd taken the picture back in 2003 because he'd never seen the green scum smother such a vast expanse of Lake Erie—it swirled from the mouth of Toledo's Maumee River out several miles into open water. "That was considered back then a big event," Bridgeman said glumly. "Now it's just average."

But why?

Lake Erie's annual average phosphorus load remains around 9,000 tons, well under its 11,000-ton target set in the 1970s and far below the 24,000 tons common in the late 1960s. Bridgeman explained that one of the problems today is rooted in the *type* of phosphorus trickling off croplands surrounding Lake Erie. Although farmers aren't necessarily sending Lake Erie any more fertilizer than they were in the days when the giant algae blooms waned, other things have changed in the algae equation—up in the sky, down on the soil and, thanks to invasive quagga mussels, throughout the waters, all in a manner that has added up to huge trouble.

It starts with the way crops are planted. Farmers have increasingly turned to no-till growing practices to prevent soil erosion. This means their fields aren't plowed over after a harvest but instead left almost as smooth as an asphalt parking lot. This is good for keeping churned-up soil from washing away in rain storms and muddying Lake Erie and its tributaries. But instead of the fertilizer being ground into turned earth, the factory-made fertilizer pebbles applied between fall harvest and spring seeding now sit like a crust on top of the fields. This is trouble if rains hit before the fertilizer has a chance to be absorbed into the crops, because it then washes away in a highly potent dissolved state.

Bridgeman explained that the difference between particulate phosphorus—the form of the mineral that caused so much trouble in the 1960s and 1970s—and this dissolved phosphorus is like the difference between a hunk of coal tossed on a campfire, and a splash of gasoline. Scientists say less than 30 percent of particulate phosphorus breaks down to a form where it can feed algae, compared to more than 90 percent for dissolved phosphorus. Studies using a radioactive marker to track the path of dissolved phosphorus through the environment have shown that within 60 seconds of its entry into a lake, half of it is absorbed by algae. Within five minutes most all of it is used.

"Dissolved reactive phosphorus is a little like money," Bridgeman explained. "It's just so valuable that it will be taken up immediately."

The problem is growing right along with the size of farms that sprawl across the old Black Swamp. Larger operations—and the equipment it takes to run them—mean fertilizer is now more commonly applied in late fall or early winter, when farmers have the free time and the ground of the old swamp land is hard enough for their heavy machinery to roll. This further lengthens the time phosphorus is prone to run off fields before the growing season—and historical data shows big early spring rains have become more common in recent years on the western end of Lake Erie, a uniquely shallow area of the lake prone to algae outbreaks known as the western basin.

Because of these changes in farming practices, the amount of the more powerful dissolved phosphorus flowing into the western basin has increased since the mid-1990s by at least 150 percent. The net result: the volume of the nutrient flowing down key rivers in western Lake Erie able to fuel algae growth is now higher than what it was when journalists were writing the lake's obituary back in the 1970s.

And when that nutrient hits Lake Erie, there are other problems in the water that weren't even imagined when the first algae reduction plans were drafted. Invasive quagga and zebra mussel numbers have exploded since the early 1990s, fundamentally altering the way life works in Lake Erie. The thumbnail-sized mollusks are what biologists refer to as "ecosystem engineers." They don't just live in the waters they invade; they actually rewire the way energy flows through them. In this case, mussels are a prime factor in the toxic algae equation because the brainless filter feeders are just smart enough to know the taste of toxic algae.

A laboratory video easily found on YouTube shows a zebra mussel in an aquarium sucking in a drifting fleck of microcystis and spitting it out with all the vigor of an unsuspecting toddler fed Brussels sprouts. Now, scale this up. There are trillions, if not quadrillions, of invasive mussels on the bottom of Lake Erie doing the same thing, around the clock, around the calendar. This incessant, selective filtering, over time, has decimated other algae populations. So, unlike the explosion of algae in the 1960s, algae outbreaks on Lake Erie today are increasingly caused by microcystis, which makes them far more dangerous.

"Lake Erie today is not the same lake it was in the '60s or '70s," explained ecologist Gary Fahnenstiel, a professor at the University of Michigan and former federal researcher who has spent much of his career studying the invasive mussels. "It does not respond to nutrients the same way."

The toxic blooms have become so regular that they can now be forecast months in advance, based largely on how much rain fell in spring, and if that rain fell before the phosphorus was absorbed into crops. In early July 2014, the National Oceanic and Atmospheric Administration predicted the annual bloom would be "significant" come late August and September.

It was still July when I drove from Bridgeman's lab over to Maumee Bay State Park next door and briefly thought of taking a dip to cool off. Until I looked at the sign on the beach that read:

Be Alert! Avoid water that:
– looks like spilled paint
– has surface scums, mats or films
– is discolored or has colored streaks
– has green globs floating below the surface.

As if the prospect of taking a swim in water choking with globs, scums, films and streaks was not enough to scare away a potential swimmer, an accompanying diamond-shaped orange sign was the deal killer:

WARNING
HIGH LEVELS OF ALGAE TOXINS HAVE BEEN DETECTED.
SWIMMING AND WADING ARE NOT RECOMMENDED FOR THE
VERY OLD, THE VERY YOUNG OR THOSE WITH COMPROMISED
IMMUNE SYSTEMS.

I opted not to risk a swim, as had, evidently, all of Toledo. The beach was deserted except for windsurf shop owner Mark Musgrave and a couple of his buddies. The water off shore was more brown than green and, despite the orange warning sign, he considered conditions to be, all in all, pretty good. Musgrave said some days the lake coughs up rotting algal mats of the spinach-like *Lyngbya* that are so thick he has to tear through them with his hands until he gets to open water. Other days, soupy blooms of microcystis give the waves green caps. "Not white caps. Green caps!" Musgrave said laughing. "It's crazy."

Those days don't stop him either. "You just live with it," he said with a smile, even if it makes some people—"the weaker ones," he joked—sick. He shook his head when I asked him if he had ever gotten ill after swimming. Then he remembered. In 2011, which happened to be the year of the record microcystis slick, he did have to take some medication.

"I had some kind of liver infection," he said, "and had to go on antibiotics."

Musgrave knows how to remedy the lake's ills, but he doesn't expect it to happen anytime soon.

"It's the farm system. You'd have to get farmers to do things differently," he said. "They'll have to get the government to step in for that."

Biologist Bridgeman echoed that point; he doesn't fault the corn, alfalfa and soybean growers whose phosphorus is conspiring with the invasive mussels to spawn the algae troubles. "They're living with the system that we've set up," he said.

The problem is they're making their living on the edge of a uniquely shallow water body (the western basin's average depth is less than 25 feet) that was once protected by the Great Black Swamp's nature-made water purification system. Bridgeman told me that local lore has it the Maumee River once ran so clear you could watch sturgeon spawning on its rocky bottom. Now it runs brown as mud—and so thick with phosphorus that it is no longer just the lake's biggest tributary but also the biggest factor in the complex algae outbreak equation. "The elephant in the room is the Maumee River, by far," he told me. "If we could work on that watershed, it would go a long way to solving the lake's problem."

But environmental regulators have their own problem; farm runoff, like ballast water, was left largely untouched by the Clean Water Act of 1972. The goal of the law was to go after polluters that own pipes, referred to in regulatory parlance as "point sources." The Clean Water Act didn't adequately tackle farm runoff, which is classified as "nonpoint" pollution. The reason, basically, is that treating what comes out of the end of a pipe is easy. Tracking and regulating what comes off farm fields is more difficult, given the vagaries of weather, soil types, field pitch and even regulators' ability to monitor what might be in the water pooling on pastures.

Change the law to force fertilizer reductions on the agriculture industry, and Bridgeman said, there is no reason the lake can't recover once again, given that its water is replaced in less than three years. But he doubted farmers would be forced to alter their ways unless something drastic occurred to wake the public up, something as dramatic as the Cuyahoga fire of 1969 that helped spur the initial phosphorus reductions. He did not know what that might be.

"It may take," he finally said with a sigh, "a major city having to shut down its water supply for a while."

There are some 11 million people who rely on Lake Erie for their drinking water. Although microcystin—the toxin produced by the microcystis—can be removed through water purification, if a plume of the tasteless, odorless stuff somehow got sucked into drinking water intakes without water treatment plant operators realizing it, it could get into the public water supply and, potentially, poison an unsuspecting population.

It would take a lot of things to go wrong for a not-so-natural disaster like that to occur, and in late July 2014 nobody was thinking it would happen anytime soon.

TWO DAYS AFTER I TALKED WITH BRIDGEMAN, I WAS SITTING IN 61-year-old farmer Norris Klump's living room on his farm west of Toledo as he tried to figure out how he and his neighbors had suddenly become such public culprits.

He explained to me that many of the farmers in the area are working the same fields their grandfathers and great-grandfathers converted from the swampland, and he said they see themselves as becoming better stewards of their inheritance with each passing year. They've stopped tilling their croplands between plantings to reduce soil erosion. They've reduced the volume of their phosphorus applications. They have embraced crop "buffers"—grass-covered strips of land along the edges of their fields meant to trap fugitive phosphorus before it washes downstream.

"We live and play and work here," said Klump, who farms about three miles north of the Ohio–Michigan border in an area his family settled in the late 1800s. "Do I really want to do something that will hurt my kids? My grandkids? Of course not. If I could do something and this will stop, yes, I would. But I don't think we're the whole problem."

Early farmers like Klump's ancestors, after all, turned the malarial lowlands west of Lake Erie into some of the most fertile croplands in the country. The farmers of today, Klump included, are proud of their ancestors' work.

"I don't think people understand the hard sweat it took to clear this land and drain it—to grow food," said Klump. "Now people are like: why are you growing stuff on this land?"

About 25 miles to the south, on the Ohio side of the state line, farmer Steve Loeffler didn't try to downplay the severity of the algae outbreaks. "It's a really serious problem," he said, noting he was particularly worried because he was planning his own vacation on the lake in a few weeks.

To do his part to stem the blooms, Loeffler explained that he plants crops of radishes that never get harvested as part of a government-funded program to soak up excess phosphorus that can then be plowed back into the soil. These cover crops also help anchor the soil when there is nothing else growing. But only about 50 acres of Loeffler's nearly 1,000-acre farm are planted with cover crops between the cash crops. "The government money to do more just isn't there," he said, adding that he'd plant more radishes on his own—if he could afford to.

"Guys can do a lot more when they're making money," he said with a chuckle.

Like Klump, Loeffler doesn't see the lake's phosphorus overload as entirely a farmer problem. "Everybody is going to have to do their part," he said. "The cities, the farms."

This is a common refrain among farmers nervous about increased fertilizer regulations, even though phosphorus measurements taken from the rivers flowing into western Lake Erie make it clear that agriculture runoff is by far the primary source of the fertilizer behind the algae outbreaks.

Loeffler pointed to leaky home septic systems. But scientists estimate such systems in Ohio are responsible for only about 88 of the more than 9,000 tons of phosphorus the lake receives annually—less than 1 percent of the total load. He also mentioned sewage overflows from cities like Toledo. But combined sewer overflows in all of Ohio are only responsible for about 90 tons of phosphorus flowing into the lake annually—another 1 percent of the total. Other farmers pointed to lawn fertilizer, though scientists say, as with detergents earlier on, phosphorus has been removed from most lawn care products even though they had never been a major factor in the algae problem. Farmers and Ohio politicians also often like to look north toward Detroit as the root of the problem.

The Detroit River does account for about 40 percent of Lake Erie's total phosphorus load. That includes wastewater from the city of Detroit as well as whatever phosphorus is sent down from Lakes Michigan, Huron and Superior. But due to the volume of the flow on the Detroit River (more than 121 billion gallons per day) and the easterly path that the current takes when it hits Lake Erie, scientists say this fast-flowing, relatively diluted phosphorus load pales in comparison to agriculture's contribution to the algae blooms. I have seen this phenomenon myself while flying over Lake Erie in late summer. Through the emerald swirls of a microcystis bloom stretching out from the mouth of the Maumee River (which has a phosphorus concentration that is about 30 times higher than the much larger Detroit River) slices a blue tongue of water that extends from the mouth of the Detroit River and bends east along the lake's northern shore.

Almost all that Maumee River phosphorus can be tracked back to farm runoff, and the farmers I talked to know things have to change. Loeffler explained to me he had cut his fertilizer applications in recent years by about half, and said a lot of nearby farmers had as well. He does it because he cares about what is going into the lake, but money is also a motivator because fertilizer costs have skyrocketed; in the 1990s the price of phosphorus-based fertilizer ran less than $200 a ton, but by 2014 it cost around $700 per ton.

Piloting a Dodge minivan down the ribbons of asphalt lacing Ottawa County farm country about an hour to the southeast of Loeffler's farm, Mike Libben, a county soil conservation program administrator, pointed out that farmers now hire soil consultants who break fields into garden-sized grids, sometimes smaller than one-tenth of an acre, and then sample those micro sections regularly for phosphorus and other fertilizer needs. "Prescriptions" are then written for chemical dosages for each grid section, often using GPS-guided equipment that allows farmers to tend to fields sprawling across hundreds of acres with the precision of a backyard tomato grower. Standing in the parking lot of a farmers' co-op outside of Oak Harbor, Ohio, Libben said the fact that the store manager had to go grab keys to show me a pile of the fertilizer illustrates just how precious the stuff has become.

"You used to not have to lock up fertilizer," he joked.

Behind that Master lock sat a small pile—maybe a couple of pickup trucks' worth—of big trouble. Libben scooped up a handful of the dusty pellets that were the size of peppercorns. The product is called 10-46-0. The numbers refer to the compound's percentage of its three key fertilizers—nitrogen, phosphorus and potassium, respectively. That means this particular pile contained almost half phosphorus—it was the strong stuff.

"I've never thought that farmers are over-applying," Libben, himself a part-time farmer with a degree in agronomy from Ohio State University, said as we gazed at the pile of pellets that looked more like cat litter than an ecosystem-scale poison, "because it's just so expensive."

Most individual farmers in the region don't think they are over-applying fertilizer any more than a laundry-washing housewife of the 1960s would have pled guilty to killing Lake Erie. A 2013 Ohio State University survey crisply illustrated this "not-my-fault" phenomenon. The vast majority of farmers in the Maumee River watershed acknowledged that farming practices were degrading water quality in the lake. Just not *their* practices. "The majority of farmers agreed that nutrient management practices improve water quality (86.4 percent) and that their own practices are sufficient to protect local water quality (76.7 percent)," stated the report.

In other words, more than three-fourths of the farmers believe they are doing their part, even as the lake is telling a different story.

Richard Thorbahn, a retired school administrator, is one of those farmers who believes he is doing his part. I visited him on a day when he was installing new drain tiles, which today aren't clay pipes but plastic tubes.

It was hot, grimy work done under a blazing sun. The ruddy-faced Thorbahn wore a long-sleeved shirt and kept his cuffs buttoned. He had on a straw hat with a handkerchief tucked in the back, Lawrence of Arabia–style. Thorbahn told me the pipe he was laying helps to alleviate the algae problems because it allows water that hits the crops to first be filtered through the soil. Water coming straight off the surface of the fields and dumping into ditches instead of flowing down to the drain tiles, he argued, is more likely to be laden with excess fertilizer. "It's surface runoff," he said, referring to farm fertilizer's path into the lake. "It's not the tile system."

Scientists disagree. They say research tracking the flow of water off fields makes it clear these drain tiles actually mainline the phosphorus into the lake, particularly when rains hit dry fields riddled with fissures and worm holes that allow the water to drop straight into the drain tiles.

Thorbahn knows as well as anybody the consequences of too much phosphorus hitting the lake. He sits on the board of the tiny Carroll Water and Sewer District, which had to shut down in fall 2013 due to a surge in microcystin.

On a tour of that water treatment facility that same week, water department superintendent Henry Biggert explained to me he wasn't particularly worried in the days before the 2013 troubles, because there was no evidence an algae bloom was approaching his water intake, about 1,000 feet off the shoreline and 30 miles east of Toledo. Biggert said his purification equipment can remove the toxin before it flows to his 2,000 customers' taps, but the protective systems have to be dialed up to do so. The warning sign he looks for is cloudy water flowing into the plant. "Usually, when you're pulling algae through the treatment plant, you know," Biggert said. "You can smell it and you're constantly cleaning equipment."

But there was no such warning when routine lab testing started showing the amount of microcystin in his plant's finished product was about three parts per billion, well above the one part per billion recommended threshold for drinking water set by the World Health Organization.

Biggert said it was likely that they had found the problem before even a drop of contaminated water had left his treatment plant, but he took the system offline for two days so the 126 miles of pipe crisscrossing the district's 26 square miles could be purged. "I couldn't live with the idea that it may be in our distribution system," he said.

It was not a big inconvenience for Biggert's customers because he was able to turn a couple of valves and switch over to a neighboring water treatment system. Even with that backup, the township afterwards went out and purchased $225,000 in new equipment to better zap the toxic algae that Biggert figured he is going to have to live with for a long time to come.

Two things about the Carroll Township shutdown were particularly unnerving. The first was that the dangerous water looked like . . . water. The toxin, which breaks loose as decaying microcystis cell walls collapse, was clear and odorless, and it was apparently drifting out in Lake Erie independent of the algae bloom that created it. The second frightening fact was how quickly Biggert's customers could have been forced to rely on bottled water to brush their teeth, wash their food and even bathe.

Biggert cringed when he thought what would happen if something similar occurred in a big city, especially if it could not just flip some switches to convert over to an emergency water source. He also worried that the toxic plumes had been growing larger and more frequent in recent years, and that at some point he may not be able to adequately treat the water, even when he knows microcystin is in it.

"The lake could get worse," he said. "And if that's the case we—and a lot of other public water systems—will have problems."

Toledo was the one city he had in mind.

THAT SAME DAY I SAT DOWN WITH ANDY MCCLURE, ADMINISTRA-tor of Toledo's Collins Park Water Treatment Plant. He was bracing for the peak algae season that was just weeks away. He tried to pull up an image on his computer to show me the National Oceanic and Atmospheric Administration website that uses satellite imagery to track Lake Erie's harmful algae blooms—"HABs" in regulatory parlance.

But every time he typed HAB, the computer jumped to contact info for Henry Biggert. The two of them, McClure said with a mild chuckle, had been communicating a lot.

McClure explained that Toledo's raw water comes from an intake about three miles off shore. The steel and concrete structure, known as the "Fortress of the Lake" when it opened in 1941, looks from afar like the prow of a battleship. It is a five-story-high cylindrical tower, on top of which sits an apartment that was once home to a full-time intake tender and his family. The apartment is long abandoned but the tower still functions as the city's water intake. Beneath it is a pipe nine feet in diameter built to suck in up to 150 million gallons of water per day from about 22 feet below the lake surface.

At the intake, potassium permanganate is added as a first-line treatment that removes odors from the water and also begins to break down microcystin if any is present as the lake water courses toward shore, a trip that can take about three hours. Once the water reaches land, it is pumped about nine miles to the treatment plant, and along the way it is further dosed with activated carbon to pull out contaminants. Alum is also added to help settle out any solids. Then the water is filtered and hit with a dash of disinfecting chlorine before it enters the pipes that feed the homes of some 500,000 Toledo area residents.

McClure told me he was confident he could clean up any microcystin-tainted water that came through his intake just by cranking up his existing treatment systems, but he worried the blooms would someday become so severe his equipment would be overwhelmed. Someday. "Things can always get worse," he told me as we sat in his office in a compound rimmed with two rows of security fencing, the taller one topped with a coiled razor wire. "And that's pretty much how you have to think."

Three days later, things got worse.

On Friday, August 1, 2014, NOAA scientists were monitoring a budding microcystis bloom at the mouth of the Maumee River. The red pixels picked up by satellite imaging signaled the first wave in the seasonal bloom lake experts had not been expecting until the end of the month. Still, nobody was worried. It wasn't a massive slick threatening to swallow the city's water intake pipe. But within hours, like a dart hitting a bull's-eye, winds drove the relatively tiny toxic plume straight into the intake. Routine sampling taken at 6:30 p.m. at the treatment plant revealed microcystin had gotten through and entered the public supply above the one part per billion threshold.

A second round of tests was ordered, and at 11 p.m. they confirmed that the toxin was indeed heading for Toledo's taps. Unlike Carroll Township, Toledo had no connection to a neighboring water supply to turn to. Toledo Mayor Michael Collins informed the Ohio Environmental Protection Agency of the crisis, which instructed him to issue the do-not-drink order. At 1:20 a.m. Saturday, August 2, a news release went out:

> DO NOT DRINK THE WATER. Alternative water should be used for drinking, making infant formula, making ice, brushing teeth and preparing food.
> DO NOT BOIL THE WATER. Boiling the water will not destroy the toxins—it will increase concentrations of the toxins.

By 2 a.m. Collins was meeting with city officials to set up a plan for how to keep a metropolitan area of a half million from panicking about losing its most basic life necessity.

"My biggest fear that evening, that early morning hour," Collins later said, was "how do I prevent panic. Because throwing panic into the equation is not going to fix the problem. It's going to increase the problem."

Nonetheless, fear spread quickly in the dark hours after the do-not-drink order went out. "By six o'clock in the morning, you couldn't find a bottle of water in any store that was open," Collins said. "I mean, none. It was gone."

Collins went on television early that morning to tell people not to panic, but that their water was safe for neither humans nor animals. He said National Guard units were rushing for Toledo from all corners of the state with pallets of bottled water and portable water treatment plants. Pre-mixed baby formula was on its way from Columbus.

The Toledo media reported that a confluence of wind, temperature and algae bloom timing created the "perfect storm" for this unnatural disaster. Those were the precise words Biggert had used when his Carroll Township water system went down less than a year earlier. Two perfect storms in 11 months.

Mayor Collins chose not to blame the weather. "Once we get our systems into the position where the water is safe again," he told local media during the height of the crisis, "that is not going to eliminate the algae problem in the western basin of Lake Erie. That is not going to eliminate the agricultural runoff."

Finally, more than two days after the shutdown, Collins appeared on television, puffy blue bags under his eyes that made him look half-raccoon, to announce lab tests had shown the water department had gotten the problem under control. He hoisted a glass of water from the city treatment plant in front of a cluster of TV cameras and took a celebratory swig.

"Here's to ya, Toledo," he said to a smattering of applause.

Nearly two months later, Collins was still rattled when he traveled to Chicago to share the lessons he had learned with fellow Great Lakes mayors, all of whom depend on the same freshwater system that provides drinking water for some 40 million people.

His fellow mayors were rapt as Collins walked them through his "nightmare."

"The canary in the coal mine is exactly what we experienced in Toledo, Ohio, on August 1st, 2nd, 3rd and 4th," he told the glum gathering. Toledo might have been the first major Great Lakes city to suffer an algae-caused drinking water shutdown, Collins warned, but it would not be the last.

"What happened in Toledo . . . it's the first time the reliability, the sustainability of our safe drinking water was threatened," Chicago Mayor Rahm Emanuel said after Collins spoke. "That moment changes this discussion."

COLLINS HOPED IN THE WEEKS AFTER THE SHUTDOWN HIS NIGHT-mare would shake the nation as the Cuyahoga fire once had, bringing demands for a solution to a problem that had been stewing for years. The year before, in fact, the nonpartisan Government Accountability Office had released a report that charged that the spectacular strides made after the Clean Water Act's passage following the Cuyahoga fire had turned into stutter steps and stumbles backward and it specifically blamed the law's failure to hold nonpoint polluters—like agriculture—accountable.

"Without changes to the Act's approach to nonpoint source pollution," the report stated, "the Act's goals are likely to remain unfulfilled."

As with the phosphorus reductions demanded of cities and industries that brought sweeping, if short-lived, algae reductions on Lake

Erie three decades ago, scientists say they know how to throttle these new toxic outbreaks. It will require a 40 percent reduction in the spring phosphorus runoffs into the Maumee River watershed. This won't completely eliminate the outbreaks, but ecologists predict it will slash them by at least 90 percent.

Don Scavia, director of the University of Michigan's Graham Sustainability Institute, notes that 40 percent also happens to be the percent of corn production in this country that goes toward ethanol. "We need," he says with an irony that could be ripped straight from the pages of Dr. Seuss, "to stop putting food in our gas tanks."

Although a recipe for a new Lake Erie fix does exist, so far the political will to execute it does not. In the months following the Toledo water shutdown, the states of Michigan and Ohio, along with the province of Ontario, agreed to work toward reducing phosphorus discharges into Lake Erie's western basin by the recommended 40 percent in the coming decade, though the agreement did not call for any new laws to actually force those cuts.

Ohio legislators took some steps of their own, banning the spreading of manure on frozen ground to keep it from flushing into the lake and requiring farmers to be trained to more safely apply store-bought fertilizer.

Collins, meanwhile, kept pushing for stiffer federal regulations to keep the next shutdown from happening, including pressing President Obama, unsuccessfully, for an executive order forcing runoff reductions. He also appeared before the U.S. Senate Agriculture, Nutrition and Forestry Committee, where he told lawmakers that he was not proud of the headlines his city had been making. But the former law enforcement officer recognized his city's problem for what it was—a Cuyahoga-like siren.

"Though as mayor I would love the pictures of the lines of those waiting for water or the images of the green algae to be forgotten," he testified, "if we forget what happened in Toledo, it is doomed to be repeated."

The next summer microcystin levels in the untreated lake water near Toledo's Lake Erie intake peaked at about five parts per billion, approaching twice the level that forced the water department to shut down in 2014. But the treatment plant operators were ready for it like firemen on a bridge waiting for a river to ignite, and they dosed the polluted water with chemicals before it could poison the city.

PART THREE

THE
FUTURE

Chapter 8

PLUGGING THE DRAIN

THE NEVER-ENDING THREAT TO SIPHON AWAY GREAT LAKES WATER

The stakes in the fight over whether to build the 1,200-mile Keystone XL pipeline to pump Alberta tar sands oil to U.S. refineries were framed in the starkest of terms. Proponents viewed the pipeline as a lifeline to a friendly oil supply for the United States, and saw the Obama Administration's denial of it as risking further entanglement in the messy politics—and mushrooming wars—of the Middle East. Opponents of the pipeline referred to the 36-inch diameter steel tube built to carry 35 million gallons per day of Canada's abundant but filthy crude oil as a "fuse" to the biggest carbon bomb on the planet.

But Canada's former U.S. ambassador saw it all as merely a prelude to future pipeline battles he predicted will be so ferocious that they will one day make the Keystone XL controversy "look silly." The reason, he said, is that the future fights will be about the one liquid that civilization literally cannot do without.

"I think five years from now we will be spending a lot of our time diplomatically and a lot of work dealing with water," Gary Doer said in 2014, at the peak of California's devastating drought.

"We have 20 percent of the fresh water (in the world) in the Great Lakes . . ." he added. "We're blessed with a lot of water, but we cannot take it for granted."

There is already a line drawn around the Canadian- and U.S.-owned Great Lakes that divides those who are allowed to tap into the world's largest freshwater system from those who are not. It's based on the natural border of the Great Lakes watershed, defined as the landmass that drains into the lakes. Water that falls inside this border—which on the landscape is often only an imperceptible bump or ridge—dribbles, trickles and rolls toward the gutters, sewer pipes and streams that feed the rivers that sustain the lakes. Water that falls just outside this line typically heads south for the Gulf of Mexico or flows north toward Canada's Hudson Bay.

Cities and towns inside the basin line are entitled to use Great Lakes water to irrigate crops, to fuel industry and to provide their residents with drinking water. Communities that lie outside the line are not. The rationale is that most all the water piped from the lakes but kept inside their basin eventually makes its way back into the lakes, often in the form of treated sewage. Water piped over the line, whether it's one mile over or 1,000 miles over, never returns, and if enough of it is diverted over time, the lakes—home to more than 80 percent of North America's surface freshwater supply—will begin to shrink.

But this haves-and-haves-not line does not hew strictly to topography. The eight Great Lakes states have recently agreed to allow water to be piped outside the basin, provided it goes to a city or town inside a *county* that lies at least partially inside the Great Lakes basin line, and provided the eight Great Lakes governors grant unanimous approval for the diversion. Canada has a parallel law.

The gerrymandering of the line was crafted by the Great Lakes states and provinces to protect massive water withdrawals from the lakes while at the same time appease those who may live only a 20 minute drive from a Great Lake shore but still live outside the Great Lakes basin—folks who, not coincidentally, tend to live in some of the wealthier and politically powerful suburbs of Great Lakes cities.

To date, this line remains for most people out of sight and out of mind, even the mind of retired schoolteacher Tom Gustafson. He lives in a Milwaukee suburb about 30 miles west of Lake Michigan in a subdivision that is so new many of its tree trunks are still only wrist thick. The village that is home to Gustafson's subdivision lies beyond the Great Lakes basin dividing line, but it is mostly inside the borders of Waukesha County, a booming county whose wells are running dangerously dry, but a county that does indeed happen to straddle the Great Lakes basin dividing line.

Yet a wedge of Gustafson's village—his wedge—happens to lie in Walworth County, which is entirely outside the Great Lakes basin. So, it appears, Gustafson's neighborhood would be therefore ineligible under the new rules to tap the lakes if and when the day comes to find a new source of drinking water.

When I knocked on his door one gray day to talk about the border he lives just beyond (it turns out that it cuts right through his next door neighbor's property), Gustafson confessed he had no idea he was living almost on top of what could become one of the 21st century's most contentious dividing lines, the manmade one that separates those who have access to the most expansive pool of freshwater on the planet, and those who don't.

"Maybe I should pay a little more attention to it," he said. "But when you think about being on city sewer and water, you just always know it's there."

Until it isn't.

The line Gustafson lives next to is going to become increasingly important in the coming years as the globe warms, precipitation patterns change and urban populations explode. Water crises, beyond the famous California drought, have in recent decades surfaced in places as close to the Great Lakes as the city of Waukesha in the heart of Waukesha County, where once-abundant groundwater supplies have been so depleted and are now so dangerously polluted with naturally occurring radium that the city is under a federal order to find a fresh, safe source for its residents. Water scarcity troubles have popped up east of the lakes in New York City, where politicians once publicly eyed the Great Lakes as a potential salve. And they have emerged south of the lakes in Atlanta, Georgia, where less than a decade ago an extreme dry spell nearly drained the public water supply and left politicians looking north for emergency relief.

Atlanta, in fact, paints a grim picture of how stretched and fragile big-city water budgets have become in the 21st century, and how far politicians are willing to go to redraw borders to access—some might say steal—another state's water. And the city of Waukesha's plan to run a pipe over to Lake Michigan to provide a fresh drinking water source for its 70,000 residents may soon reveal how well this line the politicians have drawn around the Great Lakes actually holds water.

WHEN TENNESSEE BECAME THE 16TH STATE ON JUNE 1, 1796, Congress declared that its southern border with Georgia would be the 35th parallel. That line remained nothing but a cartographical abstraction until the early summer of 1818 when two survey teams—one from Georgia and one from Tennessee, and neither, at least by today's standards, particularly good at their jobs—converged near the southern bank of the Tennessee River to divvy up what was then the western wilderness.

This was not easy work. Surveyors of that era weren't the satellite-guided technicians that they are today as much as they were artists who tried to draw lines on the ground using a clock, compass and sextant to pinpoint their positions based on celestial charts that mark stars' locations at precise times. Done right, these early 19th-century surveyors could drop bull's-eyes from the heavens and draw rail-straight lines between them. This particular effort missed its mark by a mile, and then some—1.1 miles, to be specific.

Surveyor James Camak, a University of Georgia mathematics professor with a chin weak as an owl's, knew almost right away he botched the job, but he insisted it wasn't his fault. He said he asked the Georgia governor to provide state-of-the-art survey tools but the governor declined. He also said he was given bad information: "The astronomical tables which I was compelled to use," he wrote several years after the survey, "were not such as I could have wished them to be."

Bungling a survey in this manner was not uncommon at the time—Tennessee's northern border, drawn by a different survey team, was similarly flubbed and is the reason why the state's northern border today abruptly cuts to the south on the state's western end. The ill-drawn Georgia state line probably would have gone down in history as just another quirk in the shaping of the states were it not for yet another survey done in 1837 about 130 miles south of the Georgia–Tennessee state line. That survey crew didn't muff any border measurements but the stakes couldn't have been higher when it made its own mistake. The survey team was given a simple directive: pick a spot in north central Georgia—any spot—suitable for a railroad hub from which future spur rail lines could spoke out across Georgia.

The surveyors found their spot, pounded their stake in the ground and marked it on the map as Mile Marker Zero. By the early 1840s, once the woods were cleared to make way for the train station, the name of the place had been upgraded to Terminus. Predictably, businesses and homes started to sprout beside the tracks, and in 1843 the settlement was incorporated and given a proper name: Marthasville. Railroad operators complained, so the story goes, that the name was too long to print on tickets, so in the mid-1840s the place was given yet another name, one that finally stuck: Atlanta.

It is hard to say for certain where the name came from. Perhaps it was based on the Greek mythological character Atalanta, which the baby name books will tell you means "immovable," a suitable description for a metropolitan area that is today home to some 5.5 million people. Or perhaps it is a derivative of the mythical city swallowed by water—Atlantis—which, it turns out, is anything but an apt name for a city that has become one of the most drought-vulnerable metropolises in North America.

This may sound unfathomable for a green pocket of the country that gets pounded with an average of nearly 50 inches of rain per year (soggy Seattle gets fewer than 40). But the important and troublesome fact for Atlanta isn't how much rain falls from the sky. It is how little of that precipitation then flows down rivulets, streams and rivers and into what is now Atlanta, which still sits atop a ridgeline, a very bad place for a major city to be. The driest spot on a roof in a rainstorm, after all, is the peak, where raindrops hit and immediately head, one direction or the other, for an eavestrough on one side or the other. Watersheds work similarly, and if you live in one of the fastest growing metro areas in the nation, one that is adding more than a million residents per decade, you'd want your city to be situated not on a ridge but near a trough, a place where waters gather and run deep and wide.

It is not a surprise that Atlanta was planted miles from any real river, given that the accidental city's fathers weren't worried about anything but smooth rail routes into and out of the region. What is surprising is Atlanta's failure over the last two centuries to figure out how to surgically repair this geographical birth defect. Today the city's water supply system is a hodgepodge of pipes and reservoirs that has proven to be inadequate when dry spells hit, particularly the drought that struck in 2007 that prompted a panicked Atlanta Mayor Shirley

Franklin to warn that Atlanta was tapped out, and that the city needed to go looking for water "beyond our borders."

The mayor's ominous proclamation sent ripples 500 miles north, where at that very moment the eight Great Lakes states were in a heated political fight to draw their own border around the Great Lakes to stop Atlanta—or any city outside the Great Lakes watershed—from tapping what looks like, on a two-dimensional map, an inexhaustible supply of freshwater.

In February 2008 I traveled to Atlanta to try to figure out how such a wet place could be in danger of going dry. My first stop was the tiny town of Orme, Tennessee, about 110 miles south of Nashville, that actually *had* run out of water. The creek that fed the town's water treatment plant had dried up. A tanker truck was shuttling water in from a town on the other side of the nearby Alabama state line, just so Orme residents could have a few hours of running water each day to bathe, cook, flush toilets and brush teeth. In just a matter of weeks, residents had gone from never thinking about water to never not thinking about it. This is exactly what Benjamin Franklin was talking about in 1746 when he warned: "When the well is dry, we know the worth of water."

Orme Mayor Tony Reames expressed the same sentiment to me, though in a slightly more Nashvillian fashion: "It's like having a good wife and she either passes away or leaves you. You don't appreciate her until she's gone."

Orme solved its problem with a two-mile pipeline that hitched it to the water supply in a neighboring town. It was a simple solution to a relatively simple problem. Orme had only 126 residents. It's a little more complicated when millions go thirsty and they don't have a neighbor's water supply to tap, at least not a friendly neighbor.

In late 2007, Georgia Governor Sonny Perdue warned that the Atlanta region was within 90 days of running out of water. Residents were ordered to stop filling swimming pools, watering lawns and running fountains. Some had resorted to capturing condensation drips from their air conditioners to try to keep their shrubs from shriveling. Kathy Nguyen, head of the water conservation department for Cobb County, northwest of Atlanta city limits, was receiving hundreds of calls a day. Some were from neighbors ratting out neighbors for surreptitiously sprinkling, others were from people furious they had to watch their beloved bluegrass lawns burn up. "I get lots of calls that start with 'bitch,'" she told me, "and end with click."

Georgia has no problem when normal rains fall, but trouble comes quickly when droughts strike, even if an Atlanta drought is a normal rainfall year for a typical city in the Midwest. Its high-on-a-ridge location conundrum is compounded by the fact that the region continues to add subdivisions and suburbs at a blistering pace. In the first decade of this century alone the region ballooned by some 1.6 million residents— the equivalent of adding a city about the size of Philadelphia or Phoenix.

Atlanta's primary water source is Lake Lanier, a federal reservoir on the Chattahoochee River about 50 miles north of downtown. It is a massive manmade lake—about 58 square miles with a maximum depth of more than 250 feet—if full, which is increasingly a big if. There are growing demands on that water from Atlanta as well as the downstream states of Alabama and Florida, which also have rights to use Lake Lanier for public water supplies, irrigation, industry, sustenance for the Gulf of Mexico's famous Apalachicola oyster beds and, not insignificantly, for cooling a nuclear power plant.

In early 2008, in the middle of a nearly two-decade-long legal fight with Florida and Alabama over how to allocate what water was still left in Lake Lanier, the Georgia legislature decided to open up a northern front in its fight for water. It proposed redrawing the botched border of 1818 so it would actually follow the 35th parallel. This, not coincidentally, would move the state line just far enough north to capture a section of the Tennessee River. That, Georgia legislators reckoned, would correct Camak's mistake and give their state legal rights to water in the massive river some 130 miles to the north. The idea was to run a pipe from Atlanta and pump up to one billion gallons of water per day from the river. This provoked snickers in Tennessee, until the resolution to claim a piece of their state passed unanimously in the Georgia legislature.

"What I thought was a joke has turned out to be rather disturbing," state representative Gary Odom told the Chattanooga, Tennessee, *Times Free Press*.

Legal experts generally agree that Georgia can at least make a legitimate claim to a 1.1-mile-wide strip of land along the border that covers more than 66 square miles and is home to more than 30,000 Tennesseans, but whether a court would be willing to redraw a state border that has been in place for nearly 200 years is another question entirely.

A return of wet weather later in 2008 kept Atlanta from running out of water that year, but Georgia has not dropped its quest to wrest a sliver of the Tennessee River from its neighbors. In 2013 the Georgia Legislature offered to scrap the overall border dispute in exchange for a 1.5-square-mile finger of unpopulated land extending up from the current Georgia border to the Tennessee River, just enough land to lay a pipeline to the water's edge. Tennessee balked, but it's probably only a matter of time before Georgia tries to take its case to the U.S. Supreme Court, which handles boundary disputes between states. It could be an ugly fight.

"We are not going to move the state line," Tennessee House Majority Leader Gerald McCormick huffed, "and we're not going to give them the Tennessee River."

Nobody is snickering anymore.

"Had this conflict arisen in earlier times or between independent nations it is the kind of conflict that 'might lead to war,'" the *Tennessee Bar Journal* wrote after the Georgia legislature announced its intentions. The war reference was pulled from a 1906 U.S. Supreme Court ruling in which the high court was asked to settle a water dispute between two states to the north, with the court noting at the time that if the fight had been between nations instead of states, it might well have been settled with guns rather than a gavel.

The justices made no mention that, not even two decades earlier, a cross-state water raid near the shores of Lake Michigan stirred a militia of thousands armed with whips, clubs, guns and even a piece of small artillery. And today, more than 100 years later, that battleground at the edge of the Great Lakes watershed is flaring again in a place that was once famous the world over for its seemingly inexhaustible supply of babbling spring waters. A place called Waukesha.

THE YEAR BEFORE THE GEORGIA RAILROAD SURVEYORS NAILED their stake in the forest that would become the heart of Atlanta, an 18-year-old boy with a beak for a nose and ears the size of saucers lighted out from upstate New York for the "Far West"—Wisconsin. Chauncey Olin's journey started in April 1836 on the southern bank of the St. Lawrence River, about 50 miles downstream from Lake Ontario. Traveling with his brother's family, the Olin clan boarded a steamboat at Ogdensburg, New York, and, after a "boisterous and disagreeable" two days of pushing up the violent river and across the eastern end of an ornery Lake Ontario, the family disembarked at Rochester, New York. The Olins had had enough of water travel at that point and, rather than continuing on through the Welland Canal's 40 wooden navigation locks and up into Lake Erie, arranged to roll west on a horse-drawn wagon. They were headed for Chicago, which only three years earlier was a town with fewer than 400 settlers. The first leg of the overland journey took them through the wheat fields east of the burgeoning commercial hub of Buffalo, which had ballooned in population almost tenfold to nearly 20,000 residents over the previous decade, following the 1825 opening of the Erie Canal.

The Olins, hugging the southern shore of Lake Erie, bounced through the ship-building village of Erie, Pennsylvania, before arriving at a well-worn outpost where fur traders had for decades swapped guns and gun powder for the pelts of the beaver, otter and muskrats that teemed on the banks of a river so twisting and winding the Indians called it the Crooked River, or Cuyahoga. Cleveland had incorporated as a city just eight weeks before the Olin family pulled into town behind their four horses. What struck young Chauncey most about the place was not Clevelanders' industriousness but the old-growth poplars with smokestack-like trunks so tall that their lowest branches were still 70 feet off the ground.

The Olins pressed onward and survived a trip over a muddy road built of logs laid cross-ways through a malaria-infested quagmire on the western end of Lake Erie that Chauncey called "the worst ague country in Christendom." When the family finally emerged on the western side of the Great Black Swamp, it paused in Toledo only long enough to cross the Maumee River. The Olins then headed for the open expanses of southern Michigan and northern Indiana.

"In Indiana, we saw our first prairie country, where we would travel for miles and miles without seeing a tree, shrub or house," Chauncey Olin recalled in a memoir published near the end of his life in 1893. "We said then to ourselves that it would be a hundred years before these large prairie wastes would be settled." When the Olins reached the southern end of Lake Michigan they found it easier to roll their wagon through the waves crashing on the surf-hardened sand than across the soft Indiana dunes. Eighteen days after their departure, the Olins arrived in Chicago. They were not impressed by the city exploding from a swamp.

"Chicago had been advertising throughout the East for two or three years, so it was much better known than any other western town. But we saw nothing that interested us," Chauncey wrote. "Most of the buildings were on stilts, and it was almost impossible to get through any of the streets with teams without carrying a rail on our backs to pry them out of the mud, for the streets were generally on a level with the water in the river."

The family headed north as fast as they could and three days later found themselves on the south bank of a Milwaukee River that was too deep to ford. They emptied their wagon so it could float, loaded their possessions onto a skiff and forced their horses to swim over to what would become, in less than three decades, the nation's 20th-largest city. But on the late May day that the Olins arrived in Milwaukee it was still more a wilderness camp than a village.

"We were in a new town with scarcely a dozen houses, but plenty of new-comers and Indians. After resting a few days, and looking around for something to turn up, we took our departure for what was then called Prairie Village, sixteen miles west," Chauncey wrote.

The Olins had literally hit the end of the road at Milwaukee. The only path west was a sketchy "trail" of markings carved on tree trunks that steered the family into a mud hole so big the horses had to be stripped of their harnesses and extricated one by one. Plodding along at barely one mile per hour, the Olins finally crested a subtle ridge about 15 miles west of the Lake Michigan shoreline. On the other side of this ridge the sun was setting on an expanse of rolling prairies and forests like nothing the Olins had ever seen.

"I thought this the most lovely sight I had ever beheld," Chauncey recalled. "The country looked more like a modern park than anything else. How beautiful to look upon. How strange. We said in our enthusiasm, who did this?"

The Olins would soon learn that the park-like landscape, which included countless Indian burial mounds in the shape of animals, burst with artesian springs that had only been discovered by whites two years prior. The place was home to a small Native American settlement, though it was likely an ancient one; the waters that gurgled from the ground around the calendar provided easy access to drinking water even in the frozen months. Just as importantly, those springs lured in massive herds of deer, as well as foxes, wolves and smaller mammals that provided Native American hunters, pouncing from ambush stands in nearby trees, a reliable year-round food source.

Hundreds of white settlers flooded behind the Olins, drawn to the fertile, well-watered lands that could be claimed for $1.25 an acre. The settlement was renamed "Waukesha" after a local Indian leader, even though only months after the Olins' arrival in spring 1836 the Native Americans who had long called the place home were already being pushed aside.

By 1840 the area had a population of more than 2,000 and a mill that was shipping 7,000 barrels of flour annually over the ridgeline and down into Milwaukee. By 1850 it had a population of nearly 20,000. Yet its mineral waters did not become well known outside southeastern Wisconsin until the arrival of an East Coast schemer named Colonel Richard Dunbar in 1868—30 years before Dr. Louis Eugene Perrier bought his own springs in the south of France.

Dunbar, as he told the story near his life's end, had fallen gravely ill with diabetes while building a railroad in Cuba. He was in his early 40s and had received a death sentence from his doctor when he took a grueling railroad trip out to Wisconsin from New York City to attend the funeral of his wife's mother. "My insatiable thirst was beyond the power of man to describe. My tongue and gums were ulcerated," Dunbar recalled in an 1880 history of Waukesha County, published by the Western Historical Company. "My bodily suffering was intense. Under these conditions I reluctantly made my way to Waukesha."

After the funeral, Dunbar, believing he had less than six weeks to live, grudgingly took a side trip into the Waukesha countryside to see some spring-fed farmland his sister-in-law had recently purchased. "When I entered the field on which the spring is located, the intolerable thirst which had so long afflicted me had nearly overpowered me, and at this time I bemoaned my imprudence in leaving the house and wished to return to slake my insatiate thirst," Dunbar recalled. "Miss Clarke remarked that there was plenty of water on the property which we were viewing. A tumbler was immediately procured. As if providentially, I went to the right spring. I drank six tumblerfuls, and felt instantly a most grateful and refreshed sensation."

Dunbar then downed another bucket's worth of the water that, he would later learn, was naturally laced with salts, magnesium and iron. Then he returned for the next three days to drink from what he believed was a magical spring, and by the time he headed home to New York he had proclaimed himself cured. When he relapsed shortly after his return, Dunbar was instructed by his doctor to send for a barrel of the miraculous Waukesha water. Worried that his bidders might fail to find the precise spring that he claimed had healed him (there were two other springs within 50 feet, dozens more scattered inside the village limits and even more in the county's unincorporated areas), Dunbar decided to travel back to Waukesha, and he didn't leave. He purchased the spring, christened it Bethesda, and started marketing the water nationally as an elixir for some of the 19th century's quackiest-sounding ailments: Bright's disease, torpid liver, dyspepsia, calculi and nervous prostration.

Rival bottlers and businesses followed, turning nameless springs into national brands like Hygeia, Crescent City, Clysmic, Arcadian and White Rock. But the springs weren't just an export; Waukesha was fast transformed from a cluster of little houses on the prairie—the Waukesha County village of Brookfield is, in fact, the 1839 birthplace of Laura Ingalls Wilder's mother, Caroline—into one of the Gilded Age's preeminent vacation destinations, the "Saratoga of the West" that drew in the ill as well as socialites seeking respite from the hot summers of the South and luminaries like Mary Todd Lincoln, Ulysses S. Grant and Richard W. Sears, founder of Sears, Roebuck & Company.

"Now began the 30 years of the Saratoga of the west," the *Milwaukee Journal* reported in a 1953 history of Waukesha's spring water heyday. "Now came the gay belles and beaux who had neither rheumatism nor gout and were only weary when they had danced all night. Now came the immense, round topped Saratoga trunks to be lugged upstairs and deposited in the extra large closets. Now came the young ladies and the matrons, the children in Fauntleroy suits and in blue sashes, the broadly smiling Negro nurses, the young gentlemen in striped blazers carrying both tennis rackets and silver cups to drink the water, not because it would cure them of something they did not have but for the flirtatious moments around the bubbling spring."

Nobody at the time worried for a minute that there might come a point when the springs would run dry. That was impossible for a water source that was believed at the time to be "sufficient to more than slake the thirst of all the inhabitants of the Union."

The trouble started when Chicago businessman James McElroy, a principal in Waukesha's Hygeia Spring who was selling his product at the time throughout Chicago in 10-gallon cans for a dollar—the equivalent of about $26 today—singlehandedly brought an end to the notion that Waukesha's waters were bottomless.

McElroy had a contract with organizers of the 1893 World's Fair to provide his spring water for the five-month extravaganza on the shore of Lake Michigan. The idea wasn't only to sell a glamour beverage to the millions expected to attend but to avert a public health disaster. This was nearly a decade before the opening of the Chicago Sanitary and Ship Canal that permanently reversed the flow of the sewage-carrying Chicago River so it flowed away from Lake Michigan, Chicago's drinking water supply. Until the river was reversed in 1900, Chicago's 1.7 million residents were routinely forced to drink tap water laced with their own feces. This wasn't just foul. It was deadly.

Chicago health records from 1891 reveal typhoid fever deaths had doubled over the previous year to 1,997, a number that gave Chicago the terrifying distinction of having the highest typhoid fever death rate of all the major cities in the United States and Europe. Experts of the time estimated only about 10 percent of those stricken with the illness, commonly spread through contaminated drinking water, died. That meant somewhere close to 20,000 Chicagoans that year were hit with a vicious microbe that chews its way through the digestive tract and into the bloodstream. The luckier victims suffered only vomiting, diarrhea, crippling fevers and, in harder cases, coma.

To assure fair visitors that they were not risking their lives by purchasing a ticket, fair organizers hired McElroy to deliver his Waukesha spring water through a pipe stretching some 100 miles down from Wisconsin.

McElroy received a quick, quiet approval from the Waukesha village board to build the pipeline, but as word of the plan spread so did worries that exporting Waukesha water on a pipeline scale would threaten both the village's water-based economy as well as its water supply.

"The new corporation to fill its contract would have to pump its spring low, and the spring is so situated that the others would feed into it and become dry," spring owner Alfred "Long" Jones told a news reporter of the time. "All Waukesha depends on is its water. Mr. McElroy would build hotels between here and Chicago, use Waukesha water and spoil the business here. All of this has enraged the citizens beyond measure."

The village board pulled its approval, but McElroy was not convinced he had legally lost his right to lay the pipeline, and quietly boarded a train on May 7, 1892, with some 300 Chicago ditch diggers and headed north for Waukesha. News of his plan to sneak into town and begin laying pipe under the cover of darkness arrived before the train, and McElroy was met by militia armed with shotguns, pistols, clubs and a cannon. Tough talk on the train station platform turned violent. The crowd, estimated by the Chicago *Daily Tribune* to number some 4,000, beat one of the Chicago crew's foremen black and blue. Another Chicago laborer nearly lost an eye after he was struck with the butt of a whip. McElroy, carrying a pistol and $8,000 in cash (the equivalent of more than $200,000 today) was whisked off to jail on charges of disturbing the peace and inciting a riot that never actually occurred.

"Every one thought that something terrible was going to happen," the Chicago *Daily Tribune* reported the next day. Waukesha mob leader "Long" Jones told the reporter if McElroy had tried to turn even a single spade of dirt the night before he likely would have been assaulted, if not murdered. "All of this looks ridiculous, perhaps, to an outsider, but to the people of Waukesha it is serious business."

McElroy eventually did get his six-and-a-quarter-inch-diameter pipeline capable of delivering to Chicago some 130,000 gallons of mineral water per day, but the water came from a different spring outside Waukesha village limits. The product, sold for a penny a glass, was never a big hit with fairgoers, probably because the expo organizers also purchased a water treatment system that provided safe Lake Michigan water for free. By the time the fair closed on October 30, 1893, there were, remarkably, no reported typhoid deaths among the 21.5 million ticket purchasers.

The 1892 fight between McElroy and Waukesha turned out to be the apex of the nation's fascination with Waukesha's springs. The city's water popularity would slip over the following decades along with the idea that the spring water could cure disease, and the fact that, as the Chicago fair proved, evolving water treatment technology could ensure lake water supplies were safe as anything bubbling from the ground. But Waukesha lost more than a reputation for having some of the world's most coveted water. Eventually, it lost its water altogether—or at least water safe enough to drink.

In one of history's great hydrologic ironies, today the Milwaukee suburb that is only a 20-minute interstate trip from the Lake Michigan shoreline has so depleted the groundwater that fed the famous springs that all but a handful have vanished. Waukesha is now forced to tap a pool of ancient water in wells stretching 2,000 feet below ground. Levels in this deep reserve have plunged some 500 feet and the water that is left comes out of the ground as a low-grade poison; it's laced with radium, a naturally occurring radioactive element that is a known carcinogen, at levels about three times above the federal limit. The city now uses a patchwork treatment system that removes the radium from water pulled from the deep wells and mixes it with the non-contaminated water that's left in the shallow aquifer. Despite the jerry-rigging, Waukesha still exceeds federal radium limits and is under a government order to find a new water source.

The obvious answer is Lake Michigan, which you can almost see from town. But the problem is that Waukesha, like Georgia, lies on the wrong side of a border—a border that, like Georgia, it is now trying to nudge.

IF YOU TAKE AN EAR-POPPING ELEVATOR RIDE 1,353 FEET TO THE observation deck of the landmark Chicago skyscraper now called the Willis Tower, but still known to most in the Midwest as the Sears Tower, you will have a hard time trying to explain to someone who has never seen a Great Lake the immenseness of the blue expanse below. About the best you can say is: "It looks like the ocean." But that doesn't capture the essential fact that the Great Lakes are not saltwater. So saying that is a bit like describing the Great Plains' amber waves of grain as looking like a desert. And even this commanding view from above doesn't convey the vastness of the trove that is the world's largest sweep of freshwater. Most all the water on the planet—some 97 percent—is, of course, saltwater—basically useless to humans as sustenance or for irrigation. The sliver of freshwater left over is mostly locked up in the polar ice caps or trapped so far underground it is inaccessible. This leaves the Great Lakes holding roughly 20 percent of the world's surface freshwater.

But that is just a number. Using words, it is impossible to even scratch at the depth of this natural bounty. Unless you're Keith Richards. "You go and look at Lake Superior, and you say, 'Look at all that water,'" Richards once mused. "And that's just the top!" Indeed. Lake Superior is in places more than a quarter mile deep.

It's commonly said that all this water is a gift from the glaciers, and therefore it is commonly held that the lakes are filled with glacial melt—a one-time benefaction from Mother Nature. This is not correct. The basins cradling the Great Lakes were indeed carved during the last ice age but today they constantly drain out toward the Atlantic, and constantly refill with precipitation and runoff from the rivers that feed them. But the fact that the lakes are perpetually being recharged does not make them invulnerable to schemers with plans to pipe their water off to drier lands.

Although few believe that a Great Lake—Superior or any of the others—could ever be sucked dry in our lifetimes, to think a giant lake's current shoreline can't be scrubbed from the map by water withdrawals is to ignore recent history. The most famous case of a giant lake being catastrophically drained is the sad story of the Aral Sea on the border of Kazakhstan and Uzbekistan, about 500 miles north of Iran. Until the Soviet Union decided barely a half century ago to divert the rivers that feed it to irrigate cotton fields, it was once the world's fourth largest lake. The cotton came; the lake literally went away. By

2007 the Aral Sea was about 10 percent of its former volume. The pace of the desiccation was so rapid that rusting boat hulls can still be found tilted on desert sands that were once lakebed.

From the Aral Sea's former shoreline, no water to be seen today, only a dusty, salty wasteland stretching to the horizon. Former *Newsweek* correspondent Peter Annin traveled to the Aral Sea a decade ago to paint the picture for North Americans of what a lost great lake might look like for his book, *Great Lakes Water Wars*, an exceptional treatise on the history and future of Great Lakes water diversions. It turns out that capturing in words what it feels like to stand in a desert that was once lakebed is just as hard as trying to convey the depth of a Great Lake while standing on its shore.

"Trying to describe what the Aral Sea is like is one of the most frustrating exercises of my journalism career. When you drive for five hours on the old seabed in a Russian jeep from the old shoreline to the new shoreline, how do you quantify that to somebody who has never been there?" Annin asked me. "How do you describe the magnitude of the problem when you stop and get out and look around in all directions of the compass and you can't see water anywhere and you know it was once 45 to 50 feet deep over your head?"

Although it's hard to fathom a scenario like that ever playing out in North America, a Great Lake doesn't have to disappear to cause immeasurable ecological and economic damage. Just lowering one of the lakes by several feet could be disastrous because the nearshore areas, which would be the areas affected by such a drop, are where wetlands purify storm runoff, where freighters dock, where people live and play, where cities draw their drinking water and where they discharge their treated wastewater.

Fears that drier and more politically powerful regions of the United States will someday come calling for Great Lakes water have roiled the region since at least the 1950s, when a Canadian scientist drew up a massive earth- and water-moving plan to build a dike about a hundred miles across James Bay (the southern appendage of Canada's giant Hudson Bay). The idea was to use the rivers feeding the bay to push out its saltwater and create a monstrous manmade lake, similar to what was accomplished in the Netherlands with the Zuiderzee in the early 20th century.

Under the Canadian plan known as the Great Recycling and Northern Development (GRAND) Canal, water in Canada's new freshwater lake could then be pumped south into Lake Huron. That, in turn, would allow Lake Superior, which currently feeds Lake Huron via the St. Marys River, to be tapped for irrigation of the Great Plains, and perhaps beyond. This $100 billion scheme to turn Lake Superior into an oversized water tank for irrigators did for a time enjoy, at least conceptually, support from both the Canadian prime minister and the premier of Quebec. It ultimately went nowhere, but then came a new plan to replenish the massive Ogallala Aquifer that was being drained by the farmers of the Great Plains.

Although surveys show that more than three-quarters of Americans don't know where their water comes from, it's a good bet that not many of them live above the fast-disappearing Ogallala Aquifer, which stretches from the Dakotas to Texas and once held a volume of water equal to Lake Huron. Like a lake, the Ogallala Aquifer has varying depths and its shallow areas are already drying up due to unsustainable withdrawals that have allowed farmers to make it rain around the clock in a land that receives only about 1.5 inches of rain per month. Engineers calculate that at the current rate of use the Ogallala Aquifer will be drained, almost Aral Sea–style, nearly 70 percent by the middle of this century.

When I took a trip to western Kansas in late 2003, the Ogallala had already disappeared right under the feet of Scott City farmer Robert Buerkle, a 77-year-old whose well had run dry about a decade earlier, something he thought never could happen. "When I was growing up, they kept telling us we didn't have to worry about water because there was more water here than we could ever use," he said. "Well, we proved them wrong, much to our sorrow."

In the 1980s the U.S. Army Corps of Engineers explored a plan to recharge the rapidly shrinking Ogallala by pumping water from the Missouri River into the aquifer's depleted zones. The fear at the time was that government engineers would then look to the Great Lakes to replace the Missouri River flows. There was never a formal proposal to do this, but a University of Michigan professor pushed around the numbers to see if it would be feasible. He calculated that the conveyance system to build about 600 miles of canal and pipes from Lake Superior to move some six billion gallons of water daily out to South Dakota would cost more than $19 billion in 1980 dollars. That figure did not include the power to lift all that water onto the High Plains, which the professor figured would require the equivalent of some seven nuclear power plants, at a cost of about $1 billion apiece.

I can almost hear the professor chuckling at the absurdity of it all as he tallied all the zeros in his price tag and contemplated construction of a fleet of new nuclear power plants at a time when Three Mile Island had just soured the nation's appetite for fission. This analysis is cited to this day as reason enough to scratch any notion of irrigating the Great Plains with Great Lakes water, let alone piping lake water over (or through) the Rocky Mountains and into the chronically parched Southwest. But just because it doesn't make economic sense to move vast volumes of water hundreds or even more than a thousand miles across a continent doesn't mean it won't ever happen. It is in fact happening right now.

In 1952 Mao Zedong made what must have sounded at the time like an innocuous observation about the nature of water in his People's Republic of China. "The south has plenty of water and the north lacks it," he said, "so if possible why not borrow some?" The problem for China, as for the United States, was that moving water from one side of the country to the other would cost untold billions of dollars to build the canals, aqueducts, power plants, pumping stations and reservoirs required to push Mother Nature around on a continental scale.

But Mao was famous for taking the long view. One of his favorite parables was that of the old man who grew tired of having to walk around a mountain range to get to the bank of a river, so he decided to move the mountain. The old man was scorned by his neighbors for embarking on a project that could never be accomplished in his lifetime. The frail man was undaunted. His muscle was patience; he figured his children would carry on the work, and so would their children, and their children, and so on, and on, and eventually the mountain would be moved and the people would enjoy direct access to the river. He had faith his descendants would live on forever, and he knew the mountain wasn't getting any bigger. So he liked his odds.

It took only a couple of generations—a half century, to be specific—but in 2002 the Chinese government broke ground on what is expected to be a $62 billion south-to-north water diversion project that has already forced the relocation of more than 300,000 locals. When completed the system is expected to have the capacity to move some 44 billion cubic meters of water annually, roughly the equivalent of two Colorado Rivers. The project calls for three separate diversion lines heading north from the Yangtze River, two of which are already moving water into Beijing and other thirsty northern cities in a manner that proves there is no physical reason water must flow downhill—in China or anywhere else.

In 2008 U.S. presidential hopeful and New Mexico governor Bill Richardson made his own Mao-like observation about the wisdom of sharing water across a continent. "Western states and Eastern states have not been talking to each other when it comes to proper use of our water resources," he said. "I want a national water policy. We need a dialogue between states to deal with issues like water conservation, water reuse technology, water delivery and water production. States like Wisconsin are awash in water."

Richardson was soon out of the race, but the freshwater disparity he fretted about has since only grown—dramatically. By early 2015 a water shortage on the Colorado River not seen since the grand federal dams were built in the middle of the 20th century was having a withering effect on desert cities as large as Las Vegas, where water cops patrolled once emerald-green subdivisions writing sprinkler tickets. At around the same time, California Governor Jerry Brown in early 2015 ordered a 25 percent cut in residential water use for the first time in state history. It is now believed that irrigation restrictions could be imposed in the not-too-distant future as groundwater reserves disappear and reservoir supplies become increasingly undependable.

"Right now the state has only about one year of water supply left in its reservoirs, and our strategic backup supply, groundwater, is rapidly disappearing," says Jay Famiglietti, a senior water scientist at NASA— the National Aeronautics and Space Administration. "California has no contingency plan for a persistent drought like this one (let alone a 20-plus-year mega-drought), except, apparently, staying in emergency mode and praying for rain."

California supplies at least 90 percent of the nation's lettuce, tomatoes, broccoli, cauliflower, avocados and countless other vegetables, fruits and nuts, so its water supplies are not just a regional concern. They are a national one, as national as the Great Lakes.

This imbalance may well force some sort of continental-scale osmosis in which abundant water supplies reach equilibrium with increasingly parched cities and agriculture enterprises. But it may not be the water that will flow; it may be people. Donald K. Carter of Pittsburgh's Carnegie Mellon University's Remaking Cities Institute predicts a day when the Sunbelt becomes known as the Droughtbelt—at a similar time and pace as the post-industrial upper Midwest morphs from the Rustbelt to what he likes to call the Waterbelt. "You can't have major cities and civilizations where there is not adequate water," he said. "You just can't."

Yet even if it never makes economic or even political sense in this century to move Great Lakes water out to the arid West, given the energy costs and the ever-improving (but costly and polluting) technology to desalinate ocean water, moving Great Lakes water to other regions of the country is a different matter altogether.

Only a handful of diversions of Great Lakes water beyond the lakes' natural watershed have been allowed, including Pleasant Prairie, Wisconsin, and Akron, Ohio. Both straddle the watershed boundary and are allowed to use lake water, provided they send the treated wastewater back into the Great Lakes watershed, which essentially engineered those water users into the Great Lakes watershed. The big diversion exception is Chicago, which draws about two billion gallons of water from Lake Michigan per day for drinking and other uses, and is not required to return a drop. The city sends its wastewater into the Chicago Sanitary and Ship Canal. It flows from there, under the power of gravity alone, into the Mississippi River and out to the Gulf of Mexico. That diversion has been the subject of lawsuits for more than a century, and the U.S. Supreme Court has capped the volume Illinois can take at its present level, though the canal, as designed, could carry vastly more water into the Mississippi basin—toward Atlanta or any number of Southern cities that may face future water shortages.

Great Lakes water could be exported to the East Coast in a similar way. After all, gravity alone carried water in the Erie Canal some 500 miles from Buffalo to New York Harbor. So it wasn't surprising when a mid-1980s drought left New York City reservoirs more than half empty, that the state's environmental commissioner said it was "inescapable" that New York would take a look at the Great Lakes as a potential future source of drinking water. The idea was scrapped after the drought lifted. But three years later, another diversion plan was floated by Illinois to triple flows through the Chicago canal to nearly six billion gallons per day (a volume that would indeed have a measurable effect on lake levels) to boost the volume of water in a drought-stricken Mississippi River, where water levels had dropped so far that more than 1,000 barges were stranded on sandbars from St. Louis to Vicksburg, Mississippi.

That plan also evaporated with the return of wetter weather, yet the lakes remained legally wide-open for the taking for more than a decade—until a prospector arrived in the late 1990s to lay claim to them.

IN THE 1980S, CONGRESS, PRESSED BY THE GREAT LAKES STATES, passed a law that gave each of the eight Great Lakes governors veto power over water diversions outside the Great Lakes basin. The previously mentioned diversion exemptions for Akron and Pleasant Prairie notwithstanding, the governors' collective reluctance to pump Great Lakes water elsewhere meant by the late 1990s most people who paid attention to the issue at all believed the threat of massive pipes or canals sucking away the Great Lakes was gone.

Then along came a businessman with a plan to sell Lake Superior water to Asia, one tanker at a time. And when the governors and Canadian premiers looked at their own laws and agreements, they realized they were probably powerless to stop it.

A decade before Georgia and Tennessee started tussling over their state line in 2008, Ontario entrepreneur John Febbraro unwittingly started his own border battle when he hatched a plan to send freighters loaded with Lake Superior water down the St. Lawrence Seaway, through the Panama Canal and across the Pacific Ocean to sell the water to thirsty Asian nations. To this day he maintains there was an altruistic component to his plan to make a personal profit by peddling something that belongs to everyone—the water of Great Lakes is held in public trust and managed by the eight states and two Canadian provinces that border them.

"I was doing it for a purpose," Febbraro told me, "and it was to provide clean water at a reasonable cost to Third World countries."

Febbraro said he got the idea after getting fed up with television advertisements asking for donations to provide relief to drought-stricken countries. He grumbled that the pitches were always for money and food. Nobody, he said, ever mentioned the need for water, even when the cameras panned landscapes so parched and cracked they looked to him like a jigsaw puzzle.

Febbraro got off his couch and went down to the local office of the Ontario Ministry of Environment and filled out a request to export about 160 million gallons of Lake Superior water annually. The plan, which apparently violated no Canadian law, didn't initially cause even a ripple of controversy in Canada or the United States and was approved after the required month-long public comment period. But when word landed in the newspapers about a month after the public comment period expired that Ontario had signed off on a profiteer's plan sell off Lake Superior water to Asia, politicians on both sides of the border were flabbergasted.

Nobody at the time was worried about the direct ecological impact Febbraro's diversion would have on Lake Superior, given the volume it involved. But politicians did fret about the precedent of opening the lakes to all manner of bulk diversions. "This is Pandora's Box," warned Michigan Congressman Bart Stupak. "We've always worried that somebody will try to divert Great Lakes water to arid regions. If we ship to Asia, what's to prevent shipping to the Southwest or Mexico? Where do you stop?"

Although the United States had passed legislation in 1986 that gave each of the eight Great Lakes governors veto authority over any request to divert even a drop of water beyond the Great Lakes basin boundaries, Canada had no such rule when Febbraro applied for his permit. It did have a rule prohibiting diversions that are, on average, greater than five million gallons *per day* beyond the basin boundary, but Febbraro was well within that limit.

His scheme not only highlighted Canada's inability to stop it, it also exposed weaknesses in Great Lakes protections on the U.S. side of the border. In response to Febbraro's permit, U.S. Great Lakes leaders took a hard look at their own diversion law and found it legally leaky, if not unconstitutional. The problem is that the governors never established criteria to weigh a diversion request, so if a jilted applicant were to press his case into court, the law would likely be scrapped as arbitrary—and the floodgates to divert Great Lakes water would swing open.

Under public pressure he never fathomed when he filled out his permit paperwork, Febbraro agreed not to pursue his export plan to give the governors and Canadian premiers time to stiffen and harmonize their diversion rules, an exercise that dragged on for years and one that some beyond the Great Lakes basin borders found preposterous, even paranoid.

"Do you know how hard it is to be ideologically so out of touch with reality that you've concluded Michigan could run out of water?" former Speaker of the House Newt Gingrich told a Michigan state chamber of commerce gathering in 2005 after he learned there was a controversy in the state about allowing bottled Great Lakes water to leave the basin. His audience chuckled. Nobody in the room that day, apparently, made mention of the Aral Sea, the Ogallala Aquifer or Waukesha's once "bottomless" mineral springs.

The process of getting eight states to agree on who should be eligible for Great Lakes water was not as simple as one might expect. It's easy to say no to a far-away state like Arizona; it's another matter to deny a city in a Great Lakes state that has a dire need for safe water.

It took a decade, but the Great Lakes states reached a deal in spring 2008, at the peak of the Georgia drought, that maintains their right to block most all diversions beyond the watershed. The agreement took the form of an eight-state compact that was signed later that year by President George W. Bush. (The Canadian provinces of Ontario and Quebec put together a parallel agreement, but it is separate from the compact because U.S. states cannot enter into treaties with foreign governments.) The diversion ban carries two primary exceptions. One is water that leaves the Great Lakes basin in containers that are 5.7 gallons or less. The other is water that leaves in a pipeline but goes only to a city that lies inside a county that straddles the Great Lakes watershed border, and can demonstrate that it has no other viable public water supply, and will agree to send its treated wastewater back to the lakes. The exemption was tailor made for Waukesha, which was not surprisingly the first city to apply for such a diversion under the new compact.

Its application was approved in a closely watched vote in 2016. Waukesha had requested permission to pump an average of 10.1 million gallons of Lake Michigan water each day over the ridge separating Lake Michigan and the Mississippi River basin—a mere hump in the landscape that Chauncey Olin probably hadn't even realized he crossed on his trip to Waukesha back in 1836. The original Waukesha request was for almost 100 times more than the volume of water proposed for the World's Fair pipeline that brought out the Waukesha militia in 1892, but it is an immeasurably small amount in the context of the Great Lakes—like a single teaspoon taken from an Olympic-sized swimming pool, not even a drop in the bucket.

Of course, the withdrawal request that prompted the creation of the new compact between the Great Lakes was described as barely a drop in the bucket.

Waukesha's push for lake water was widely blasted as stretching terms of the compact. Opponents argued that the city could fix its radium problem by adequately treating—and conserving—the ground water it still has. Furthermore, although Waukesha has agreed to send its treated wastewater back to Lake Michigan as part of a $207 million pipeline plan, it won permission to receive more water than the seven million gallons per day it presently uses. It isn't about the volume of water this one city is requesting, critics say; it's the precedent it might set if the tenets of the compact are not strictly enforced.

"We are not opposed to a Waukesha diversion. We are opposed to a diversion that does not meet the standards of the compact," Marc Smith of the National Wildlife Federation told me in the days after the Waukesha application was formally filed with the Great Lakes governors. "We want to make sure the integrity of the compact is upheld. We want to make sure it works."

What he and other early opponents of Waukesha's plan wanted was to make sure the new line around the Great Lakes would hold, as Waukesha's probe of it is certain to be just the first in the coming years and decades—and beyond. Because as long as the only thing standing between a thirsty population and a new water source is a boundary that the dry side sees as improperly drawn, there will always be a case to be made to move it, even just a little bit, even if that line has been in place for centuries. Just ask Tennessee.

Chapter 9

A SHAKY BALANCING ACT

CLIMATE CHANGE AND THE FALL AND RISE OF THE LAKES

P atric Kuptz knows Lake Michigan like few others. "I've spent most of my life within 50 feet of here," the 37-year-old Milwaukee native and marina worker told me on a sunny May morning in 2013 as he worked on a sailboat near his third-generation family home, a brown brick duplex at the edge of the city's South Shore Yacht Club. This is where Kuptz spent his boyhood summers chasing perch lurking in the pale green water around the docks, splashing in the frigid surf and making all manner of mischief around the yacht club, news of which often made it home to his parents before he did.

But by early 2013 Kuptz hardly recognized the lakeshore as the one he grew up on. He pointed to a sandy beach that didn't even exist when he was a kid in the mid-1980s, when the water hit a record high level— some six feet higher than it was on this morning—and yacht club members needed homemade wooden steps to ascend from the docks to their boats. When that high water dropped a couple of years later and those stairs were being thrown away, the young Kuptz couldn't believe the adults' shortsightedness. He knew even as a child that Great Lakes water levels are a fickle, ever-fluctuating thing, varying by about a foot between summertime high and wintertime low, and by as much as several feet over a period of years due to long-term weather patterns. The young Kuptz was convinced it was just a matter of time until those high water levels returned, so he hatched a plan to stash the stairs in his garage and sell them back to their owners when the lake bounced back.

He never acted on that plan, and it was a good thing because the only record water levels to return since was the record low Lake Michigan hit in 2013, and that was after going the better part of two decades (another record) without climbing back to its long-term average. Now it was the adult Kuptz who had become convinced that the lake wasn't coming back, at least not in his lifetime, and he was the one making plans accordingly. He sold his own sailboat.

"I actually bought a power boat because I'm worried about the draft," Kuptz said of the keel-scraping waters plaguing the harbor. "It's nuts. I'd never seen it this low." Nobody had.

PREHISTORIC WATER LEVEL RECORDS, RE-CREATED BY RADIO- carbon dating the crusty ridges rimming the lakes that are the remains of ancient beaches, reveal that about 4,500 years ago Lake Michigan was roughly 13 feet higher than it is today. Then, likely driven at least partially by drought, it underwent a steep decline over a 500-year period before settling near levels closer to what we've known.

Some people might take comfort in the notion the lake has previously experienced swings that are even greater than those that are rattling people today. But it's important to remember that Lake Michigan's 13-foot plunge is ancient history. It happened before the existence of lakeside cities bursting with millions of people and trillions of dollars of homes, skyscrapers, factories, rail yards, roadways, navigation channels and canals, sewage treatment plants, drinking water intake pipes and nuclear reactors, all clinging to the modern shoreline, all needing the water levels to stay basically where they have been for the last century and a half.

But this is becoming increasingly unlikely as the effects of climate change take hold on what might prove to be one of the most temperature-sensitive regions on the globe. Most scientists agree that warmer water and air temperatures will super-charge the Great Lakes' precipitation–evaporation cycle; more water will go up into the sky, and more will come down. The question is whether the cycles will balance each other out. For most of the past two decades they have not.

On one hand then, Great Lakes water levels fluctuate constantly due to seasonal and long-term weather patterns. But the lakes are also, essentially, a giant slow-motion river that flows from the middle of the continent out to the North Atlantic—and not entirely naturally.

Lake Superior, which sits at the top of the system, drifts oh so slowly, toward a gap on its eastern shore that creates the headwaters of the St. Marys River, which takes from Lake Superior an average of about 75,000 cubic feet of water per second. The St. Marys tumbles south into Lake Huron, whose outflow is the Lake Erie–bound St. Clair River, which has an average flow of about 181,000 cubic feet per second. Lake Erie tilts into the Niagara River, which thunders over Niagara Falls and into Lake Ontario, which pours into the Atlantic Ocean–bound St. Lawrence River some 251,000 cubic feet per second—that's roughly enough to provide for all the drinking water and agriculture needs of the entire United States.

There are two manmade control points on the system that can slow and speed flows out of the lakes. A dam on the St. Marys River that can be operated to tweak Lake Superior's levels by mere inches is part of a hydropower and navigational lock system that allows freighters to sail up into Lake Superior from Lake Huron. There is also a dam system on the St. Lawrence River below Lake Ontario that allows for some degree of control over Ontario's level. There are no human controls on the St.

Clair River that drains Lakes Michigan and Huron, which are actually one grand lake connected at the Straits of Mackinac.

There are also three primary manmade diversions that allow manipulation of the amount of water entering and leaving the Great Lakes, the most notorious of which is the Chicago Sanitary and Ship Canal. But this diversion of Great Lakes water into the Mississippi River basin is capped by the U.S. Supreme Court and is more than compensated for by two diversions *into* the Great Lakes; Canada channels two rivers out of the Hudson Bay watershed and into Lake Superior. Those were World War II–era projects that Canada undertook in order to squeeze more hydroelectric power out of the lakes' rumble to the sea.

Yet even with all this tinkering, Michigan and Huron, since record keeping began in the 1800s, had always stayed within about three feet of their long-term average, and did so almost rhythmically. Previous drops into low water, in the 1920s, '30s, '50s and '60s, were always followed by a quick and sustained rebound back to—and then beyond—the long-term average. Usually it happened within three or four years, though the slow but steady climb during the Dust Bowl droughts took the better part of the '30s.

And then came a crash no one saw coming. Between 1998 and 1999, Lakes Michigan and Huron took a three-foot plunge—a record drop from one year to the next—and nearly surpassed the all-time low set in the 1960s. Lakes Michigan and Huron then went a record 15 years without climbing back to their long-term average. Then in 2013 Michigan and Huron hit a low never seen in a century and a half of record keeping—about 6.5 feet below the high water period of the 1980s. Hydrologists were flummoxed. Unlike previous low-water eras, this one could not be blamed on a prolonged drought. The decade and a half that Lakes Michigan and Huron remained below their long-term average had actually been, on average, *wetter* than usual.

So where did all that lake water go? This is not a story about climate change. It's a story about climate changed.

AFTER WEEKS OF PATIENTLY WAITING FOR ICE TO SLOWLY BUILD on Lake Superior, Bob Krumenaker finally decided on a chilly St. Patrick's Day in 2013 to click into his cross country skis and venture out to one of the islands he oversees as superintendent of Apostle Islands National Lakeshore. The ice felt solid as concrete when Krumenaker and a colleague glided their way out to Basswood Island, about a mile offshore. A little further out, the lake was dotted with ice fishing shanties and snowmobilers and it all appeared that the ornery water had once again lapsed into a deep wintertime slumber under a blanket of ice as durable as any ribbon of interstate. This was a mirage.

The reality was hidden under Krumenaker's green pullover. He wore a red life preserver. The frozen white coastline belied what was happening just a few miles out, where the ice gave way to the churning black waves of a body of water that never went to sleep for the winter. Although Krumenaker said he never feared for his safety, it was the first time he wore a life jacket while on skis. His wife insisted. Several weeks earlier, not far from Krumenaker's park headquarters in Bayfield, a fishing guide who was a former police officer had plunged through the ice and died. Less than a month later, two more Bayfield-area snowmobilers crashed through the ice off nearby Madeline Island and died. "Three local people who are familiar with conditions dying," Krumenaker ruminated several days after his ski trip, "has definitely rattled this community."

A satellite image of the ice taken in the days after that ski trip—a time of year when the percent of Lake Superior covered by ice is usually at its peak—revealed that the lake was only frozen over in the harbors and bays of cities like Bayfield, Duluth, Copper Harbor and Thunder Bay. The pictures taken from space revealed that the rest of Lake Superior—about 90 percent of its surface area—was open water, black as the type on this page. Researchers who have been tracking Lake Superior's ice cover for decades with airplane surveys and satellite imagery say this is now normal.

Historically, on average, about a quarter to a third of the surface of Lake Superior, an area of lake roughly the size of Massachusetts, froze each year. But average ice cover for Superior declined by 76 percent between 1973 and 2011. A similar phenomenon has occurred across the Great Lakes region; one federal research estimated in 2013 there had been a 63 percent drop in average ice cover for all the lakes over the past four decades. Across the same time period, scientists calculate there has been a mere 1.6°F upturn in the over-water air temperature for all the Great Lakes—with most of the change happening since the late '90s. Common sense says the stubbornly frigid inland seas would be immune to such a subtle bump in air temperature. It took a scientist from outside the Great Lakes to figure out just the opposite is true.

Jay Austin arrived at the University of Minnesota–Duluth in 2005 with a doctorate in oceanography from the Massachusetts Institute of Technology–Woods Hole Oceanographic Institute, but almost no experience studying freshwater bodies. "I didn't know anything about lake temperatures so I thought a good way to learn would be to grab the data, plot it up and play with it for a week and see what happens," Austin told me in the hallway of his office building perched on a San Francisco–steep Duluth hillside on the western end of Lake Superior. It was spring break 2013 and all his students were gone, but there was nothing spring-like about this particular morning. Austin was wearing a turtleneck and ski sweater, and the lake below was pounded by a late-season snow squall that masked, albeit temporarily, the profound temperature and hydrological changes he has helped to uncover.

Working with a century's worth of water temperature records harvested from a monitoring station on the lake's eastern end along with
data retrieved from a network of mid-lake weather buoys, Austin found
something that made little sense. Since 1980, Lake Superior's average
summer surface water temperature has been climbing at a rate of about
two degrees per decade, roughly double the rate of the increase in air
temperature over the Superior basin.

Austin went looking for the scientific literature explaining how
this could be. He found almost nothing. He eventually went to the ice-
cover data and, by comparing historic water temperatures to ice records,
determined it's not just warm summer weather driving the increase in
water temperatures. It's also what's happening in winter. The air temperature increase in the Great Lakes region, however slight, has been
enough to dramatically reduce Superior's average ice cover. And without a bright white winter cap to bounce solar radiation back into the
sky, the lake continues to soak up heat, even during the snow season.
It turns out this jump-start on the annual warming process has a profound effect on peak surface water temperatures during the summer.

"The intuition is that a very large lake like this would be slow to
respond somehow to climate change," he said. "But in fact we're finding that it's particularly sensitive."

The outsized role winter ice has on Superior's summertime water
temperatures jumped out of Austin's data. The more ice coverage in
winter, the cooler the lake is likely to be the following summer; the
less ice, the warmer the summer water. "It's not one of those things
that you had to do a lot of statistical analysis to convince yourself of the
significance," Austin said. "You made the plot, and there it was."

A similar phenomenon is under way on the other Great Lakes.
A weather buoy in southern Lake Michigan showed a 3.4°F degree
increase in average summer surface water temperatures between 1997
and 2013. One day in 2012 that mid-lake thermometer station, located

about 43 miles southeast of Milwaukee, recorded a Caribbean-like 80 degrees. It was only the beginning of July, a time of year when the lake temperature is more likely to be in the high 50s. "There has been a change in air temperatures. It's not dramatic, but it's just enough to not produce the ice coverage we used to have," said Paul Roebber, a University of Wisconsin–Milwaukee meteorologist. "And that makes all the difference in a system like this."

Warmer water might not sound like the troublesome news that it is; it obviously makes for friendlier lakes for beachgoers and tourists. Superior's temperature bump, for example, means on hot summer nights oceanographer Austin can take his boy swimming at a beach near downtown Duluth, something most people would never do a generation ago because the lake stayed so cold year-round that even summer swimmers risked hypothermia.

But the temperature increase could prove disastrous for the lakes' long-term water levels because it is driving up evaporation rates. NOAA data show that evaporation on Michigan and Huron was above average every year from 2013 dating back to 1999, when the lakes' record long low-water era began. With little to no protective winter ice cap, chilled air whooshing over relatively warm water leads to more evaporation. The result of this thermal avalanche triggered by just a tiny blip in air temperatures: the surface of the lakes is going poof into the sky.

Although increased evaporation happens when cool winds blow over the summer's ever-warming waters, the biggest change has come during the frigid gales of October, November and December, a time of year when evaporation can sap up to two inches of water *per week* from the still relatively warm lakes. Roebber pointed to data analyzing how much water Michigan and Huron have lost to evaporation each year between 1997 and 2013 compared to the historic annual average. Tally all those numbers, and it showed that increased evaporation, calculated by a model using water temperature, air temperature and wind speed, had sucked more than four feet from the lakes. Some of that above-normal evaporation loss was offset by above-normal precipitation, but Roebber said people need to understand something has fundamentally changed in the lakes' historic low-water, high-water cycles. He is not alone.

Frank Quinn, who retired as a hydrologist with the National Oceanic and Atmospheric Administration's Great Lakes Environmental Research Laboratory in Ann Arbor, Michigan, has tracked lake level fluctuations for more than a half century, and although people have long grumbled about the episodes of high water, he says they rarely blamed anyone but Mother Nature. As for low water, Quinn heard all manner of crazy theories. The evaporative effect of atom bomb testing was a popular culprit in the 1960s, and rumors swirled for years of a secret canal under Niagara Falls channeling flows toward the arid West. The truth was always a little drier—the lakes were simply suffering from a lack of rain and snow. Or at least that's how things used to work.

"What appears to have happened is the hydrologic regime—the climate—has changed," Quinn said in 2013. "We're getting the precipitation, but we're losing a lot more water through evaporation and that is what is causing the drop in Lake Michigan and Huron's water levels—the continuing low levels, the part we can't explain."

Quinn wasn't saying a change like this hadn't happened before humans started taking meticulous notes of bouncing water levels, nor was he saying that at some point it wouldn't be reversed. But he said there is no question that during the first decade of this century there was a profound switch in the way the lakes work—to the point that in some years the sky actually saps more from the lakes than it gives.

But this change in the evaporation–precipitation cycle was not the only reason Lakes Michigan and Huron plunged into uncharted territory in 2013. Part of the problem can also be found at the bottom of the St. Clair River, which is, essentially, a drain hole for Lakes Michigan and Huron on their rush toward the ocean—a drain hole that has been ravaged for the sake of navigation.

IT WAS DEAD CALM AT THE SOUTHERN TIP OF LAKE HURON ON August 4, 1900, when, at just about midnight, the 231-foot-long *Fontana*, loaded with 2,600 tons of iron ore and southbound for the steel mills of Cleveland, neared the most dramatic hydrological chokepoint in all of the Great Lakes—the 800-foot-wide flume that is the mouth of the St. Clair River that carries the collective outflows of Lakes Superior, Michigan and Huron.

As the *Fontana* approached this roiling hole at the bottom of Lake Huron, out of a thin fog came the lights of the upstream-bound, 324-foot-long *Santiago*. Both the *Fontana* and *Santiago* were sailing vessels that on this night had been converted into barges, and both were being towed about 800 feet behind modern steamships. The captains of the two steamers greeted each other with horn blasts when their distance closed to about a mile, and when their paths crossed in the rushing river channel they steamed past each other without incident. But then the *Santiago* whipped wildly on its tow line in the swirling, churning current and veered toward the *Fontana*, cracking open its hull and instantly sinking the ship in one of the quirkiest maritime disasters in the history of the Great Lakes, one whose consequences are still felt—for good and ill—today.

The *Fontana* and one of its crewmembers were lost in the collision. The bigger disaster was that the ship foundered in the worst of all possible places—the middle of an excruciatingly tight and treacherous shipping channel with a current ripping along at about five miles per hour—faster than the Colorado River's plunge through the Grand

Canyon. This posed an obvious risk to all the other vessels trying to weave their way through the zone known as "the rapids," a remarkable moniker for a stretch of river that runs up to 70 feet deep.

About three weeks later, before anyone could figure out how to dispose of the wreck, an upbound steel-hulled steamer trying to squeeze past the remains of the *Fontana* collided with the downbound schooner-barge *John Martin* with such violence that the splintering of the *John Martin's* timbers could be heard on land a mile away. Four of the *John Martin's* crewmen drowned as the river sucked the 220-foot-long vessel to the bottom in less than 20 seconds. The U.S. government eventually removed all the pieces of the two boats that jutted high enough into the shipping channel to pose a threat to the freighters passing above, though the hulls of both remain on the river bottom to this day.

The loss of two ships and five lives on a set of inland seas that have swallowed some 6,000 boats and 30,000 souls would, more than 100 years later, typically merit only a couple of lines on a maritime historian's spread sheet. But there was something remarkable about these two shipwrecks—the boats fell in such a manner that they have throttled the natural flow of the St. Clair River enough to raise water levels on Lakes Michigan and Huron by more than an inch. This is a lot of water. Lakes Huron and Michigan sprawl across an area the size of Maryland, Vermont, New Hampshire, New Jersey and Connecticut. Another way to look at it: the Chicago canal is 160 feet wide and 25 feet deep and sucks more than two billion gallons a day away from Lake Michigan, and this has lowered Michigan and Huron's long-term average level by a mere two inches (a loss more than compensated for by the two Canadian diversions *into* the Great Lakes).

The fact that the relatively tiny *Fontana* and *John Martin*, carrying only a combined 4,200 tons of cargo, could so cork the primary outflow for Lakes Michigan and Huron should have been a warning to the U.S. and Canadian governments: don't fiddle with the St. Clair River if you don't want to mess with water levels on the two Great Lakes that feed it. They didn't heed it.

In the three decades that followed, more than three million tons of sand and gravel were mined from the St. Clair riverbed, enough material to fill some 300,000 dump trucks. If you think of the St. Clair River as the drain hole in the giant bathtub made up of Lakes Michigan and Huron, then this work was like taking a jackhammer to that drain. Hydrologists of the time knew all the excavation had boosted flows on the St. Clair River in a manner that was sucking down the lakes that flow into it, and riverbed mining was finally banned in 1926. But navigational dredging to allow ever-bigger freighters to sail up from Lake Erie continued on a scale that's hard to grasp. One dredging project in the 1930s scooped up 11 million cubic yards of riverbed. Another round of dredging in the early 1960s to deepen the 39-mile-long river's navigation channel to match that of the new St. Lawrence Seaway removed another 2 million cubic yards. All told, starting with the first scraping of the shallowest chokepoints in the 1850s to allow schooners to navigate the river, engineers estimate more than 33 million cubic yards of riverbed have been removed from the St. Clair River's shipping channel.

The engineers in charge of the 20th-century navigational dredging were aware that all their work was dropping the long-term average levels of Michigan and Huron—and not in a manner measured in mere inches. That's why each time the federal government approved a major St. Clair River dredging project over the past century it came with a plan to compensate for all the water lost to the sea. The idea was to build dam-like structures in strategic areas on the river bottom to slow the flow but still maintain a channel deep enough to keep the Great Lakes freighter business afloat. Yet that compensation work was not done in the early 1900s when the St. Clair River's shipping channel was deepened to 22 feet, nor was it done following the 1930s dredging that took the channel to 25 feet. And it wasn't done in the early 1960s, when the channel was excavated to roughly 30 feet. Two years after that last dredging project, Lakes Michigan and Huron plunged to what was at the time an all-time low. This was not a big surprise; hydrologists calculated all the human dredging and mining of the St. Clair River up to that point had collectively lowered Michigan's and Huron's long-term average by as much as a foot and a half—enough water to turn a landmass the size of Denmark into a swimming pool.

This "lost water" is a bit of an abstraction because, of course, Great Lakes water levels are in constant flux. Plotting a lake's level over a period of decades onto a graph is like looking at an EKG printout. Blips and dips reflect seasonal oscillations and the larger undulations span years and even decades. If you draw a line through the middle of those highs and lows, you get a lake's long-term average surface level, and it's that straight line for Michigan and Huron that has been lowered by the St. Clair dredging. What it means is that if you walk down to the shore of Michigan or Huron today—or any day—the water is about a foot and a half lower than it would otherwise be had humans not meddled with the river bottom.

Although no construction project to compensate for all this dredging damage was ever actually done, hydrologist Quinn was part of a team the U.S. Army Corps of Engineers assembled in the 1960s at a sprawling federal research center in Vicksburg, Mississippi, to explore how it might be accomplished. The team built a miniaturized southern Lake Huron and St. Clair River. The model of the river's upper 2.8 miles was made of concrete and meticulously sculpted based on aerial photos and bathymetrical surveys. The engineers further calibrated their three-dimensional replica of the river by dropping floats in the real St. Clair and tracking their path and the time they took to cover it. Then they tweaked the flow and shape of the model river accordingly to match the real river's depth, velocity and flow paths. The engineers then built a precisely shrunken, remote controlled scale model of a 730-foot bulk carrier to ensure they got flows just right—not only for the lakes but for the shipping industry the Army Corps served. Then the engineers, all former boys, added some dioramic flourishes. The mouth of the river was spanned by a scaled-down version of the bridge that connects Port Huron, Michigan, with Point Edward, Ontario. On the riverbank was a miniaturized Coast Guard station, which could have had nothing to do with replicating river hydrology but certainly must've added a flair to the whole production.

I'd heard references to this incredible model for several years, and when I visited the now-retired Quinn at his home outside of Ann Arbor, Michigan, I asked him if it might still exist. Perhaps it was stashed somewhere in a warehouse? Quinn chuckled. The long-discarded model, he told me, *was* basically a warehouse; the river portion alone sprawled nearly the length of a football field, and the model boat, apparently lost to the ages, was almost 12 feet long.

Once the river and boat were operational, the next step was to install the model underwater berms in a precise manner that would throttle the flow of the water but still allow the freighter to pass above or around them. The engineers zeroed in on the exact places to install the structures, and even used crushed coal to simulate how Lake Huron sediment tumbling into the river might at some point affect the shape of the berms which could, in turn, alter their effect on river currents and lake levels. By the time the men were done with their studies they knew with mind-boggling accuracy how to build a system of underwater dams in a manner that they could peg the resulting rise on the long-term average levels on the real Lakes Michigan and Huron to within six-tenths of an inch.

"They did it right to scale, and then put water down the river to see if it would work and how it would work," said Quinn. "It worked just fine."

But the Army Corps never built the real thing.

"What happened then was, starting in about 1966 or so, we started getting a lot of rain and the water levels went up," said Quinn. As lake levels climbed back toward their long-term average, interest in the project evaporated. So the scale model of the river was junked. All that remains today is a dusty plan—a plan that might never have been shelved had the modelers gone a step further.

"It should be remembered that the model was of the fixed-bed type and was not free to erode by action of the currents," the engineers wrote in their final report. "In these tests, the bed of the river was assumed to be reasonably stable."

This was a big assumption.

LAKE HURON COTTAGE OWNER MARY MUTER NEVER BELIEVED the recent record string of low water years was entirely the result of rising water temperatures. The retired nurse from Toronto acknowledges that unprecedented evaporation did contribute to the record low water levels on Michigan and Huron in 2013, but she also blames unforeseen consequences of the St. Clair dredging, and what she sees as the government's "irresponsible" decision to never compensate for it. Muter owns a home on Lake Huron's Georgian Bay, one of the most vulnerable areas in all the Great Lakes to low water because of the sloping nature of its shoreline. When water levels drop a foot or two in island-freckled Georgian Bay, it doesn't just mean more air between boat docks and the water. In some places it means the water pulls back more than 100 feet from the old shoreline, sometimes making island properties that are only accessible by boat virtually unreachable.

Muter had been eyeing the St. Clair River with some suspicion after Huron's mysterious three-foot plunge in 1999. She had a vague understanding of the river's dredging history and the critical role the size of the river channel plays in governing lake levels. This made her wonder if something more was at play than just a change in weather patterns, and she finally decided to travel south to see the St. Clair's headwaters for herself. Muter is not a hydrologist and she already viewed the river with some suspicion. So when she got out of her black Volvo that day back in 2001 in the gritty oil refinery city of Sarnia, Ontario, it was probably no surprise that she found herself stunned at the pace of the river current, against which small boats struggled upstream into Lake Huron, like birds flying head-on into a gust.

Her surprise that the sandy-bottomed river was only a couple of feet deep several feet off shore turned to amazement when she looked up and saw a battleship-sized freighter gliding by impossibly close. "I said to myself: how could it get that deep, from where I'm standing to where that ship was," Muter recalled. "The ship was only 100 to 150 feet from me." She started digging into the historical dredging records and the never-finished plans to compensate for the lost water. Then she helped spearhead a $250,000 fundraiser to hire an engineering firm to get to the bottom of what was going on in the river. She got the answer she was looking for in 2004, when her group released an alarming study that said unexpected erosion since the 1960s had lowered the lakes well beyond the 14–18 inches lost from the lakes due to riverbed dredging and mining that the U.S. and Canadian governments had previously acknowledged.

The Georgian Bay study claimed that the total loss tied to riverbed mining, dredging and subsequent erosion was actually more than two feet—and getting worse by the day. It theorized that the 1960s dredging had indeed scraped away a durable layer of cobble and rock, exposing soft riverbed that began to erode in the violent current, leading to ever-increasing water loss. The engineer who conducted the study explained that the problem was compounded by development along southern Lake Huron's shore and along the St. Clair riverbank in recent decades that blocked sand and other material from flowing into the river channel and filling in eroded zones.

The gist of the Georgian Bay report: the rock-solid plug that had kept Lakes Michigan and Huron in place for thousands of years had been turned to mush.

The Army Corps acknowledged that something indeed was amiss, pointing to its own numbers tracking the relative surface levels of Lakes Michigan and Huron compared to downstream Lake Erie. Because the two systems are connected by the St. Clair River, when Michigan and Huron drop, Lake Erie historically also dropped. But in recent decades the approximately nine-foot difference in "head" between the two bodies of water has been shrinking.

The Army Corps wasn't convinced this was necessarily a St. Clair River erosion problem. Other plausible explanations for the shrinking difference between Michigan–Huron and Lake Erie include shifting weather patterns that sent increased precipitation over Lake Erie, as well as the ongoing, geologically slow rebound of the earth's crust after being compressed by the last glaciers: it has been well documented that the Georgian Bay region is inexorably rising in relation to the southern shores of Lakes Michigan and Huron in a manner that, in effect, simultaneously lowers water levels in Georgian Bay and raises water in areas to the south. The Georgian Bay group acknowledged these factors as likely contributing to the shrinking difference between the levels of Michigan–Huron and Lake Erie, but it maintained the biggest problem was the Army Corps' dredging that accidentally created an ever-eroding St. Clair river bottom. "We've got something alarming going on here," said the engineer who conducted the Georgian Bay study.

The news was alarming enough for the International Joint Commission, the bi-national board that oversees U.S. and Canadian boundary waters issues, to hire a team of scientists to look into the question. In 2009 those scientists, led by a career Army Corps employee, released a report that concluded unexpected erosion since the 1960s had indeed lowered Michigan and Huron by an additional 3 to 5 inches, though they said the erosion appeared to stop somewhere around 1999, and therefore the water loss problem was not worsening. This means the official acknowledged toll on Lakes Michigan and Huron tied to St. Clair River mining, dredging, coupled with erosion since the early '60s, is as much as 21 inches—nearly two feet, a figure many in Georgian Bay believe is still a vast underestimate of the impact of erosion, which they also argue is ongoing.

The scientists who conducted the study for the Joint Commission recommended that no riverbed restoration project be ordered to fix the problem and instead advised that the region just learn to live with the lost water. The Joint Commission, bitterly split on the issue, voted to reject the advice of its scientists and has instead asked the U.S. and Canadian governments to explore, once again, water-slowing contraptions on the St. Clair.

Whether to continue tinkering in the St. Clair River in the 21st century, this time not to cater to the shipping industry but to undo the dredging damage done in the 20th century, is becoming every bit as contentious as Ontario's John Febbraro's plan to ship off tankers full of Lake Superior water to Asia. Lana Pollack, the U.S. chairman of the Joint Commission, refused to sign the letter from the commission recommending that the U.S. and Canadian governments explore what it will take to bring lake levels up. She told me she fears any such restoration project only offers "false hope" and distracts the public's attention from what she sees as the real issue—climate change causing the increased evaporation.

"Some of the very same people who deny the reality of climate change being caused by our energy choices are the same people who say, 'We want you to fix this,'" she said. "So on the one hand they say mankind is too small to impact Mother Nature—that forces of nature are much stronger than the impacts of man. Yet they somehow turn around and say, 'OK, governments: put a plug in—engineer something, dredge something, dig out, blow up, modify.' They don't think man is too weak to engineer a fix, but they somehow say we're not responsible for the cause."

Just after the lakes hit their nadir in 2013, Tim Eder, the executive director of the Great Lakes Commission, an agency created by the Great Lakes states to protect the ecological integrity of the lakes while squeezing as much out of them as is economically possible, was of a mixed mind on whether to slow the St. Clair flow to raise the level of Michigan and Huron. It could be good for navigation, for recreational boating and for property owners. But he saw potential downsides, the most obvious of which is that the lake levels could rise naturally. A river restoration would then exacerbate flooding and erosion in heavily populated places such as Milwaukee and Chicago. He also worried that if the Army Corps were tasked with building and operating some kind of controlling mechanism on the St. Clair River, people would then demand the lakes be managed to eliminate or reduce their natural fluctuations, which are critical to sustaining near-shore fish habitat as well as maintaining the health of wetlands and marshes; lake levels, after all, are nature-built to fluctuate—to a degree.

The Joint Commission did not recommend the governments explore a dam with controllable gates on the St. Clair to control flows, only that they erect the berms the Army Corps had previously considered. This would raise lake levels but still allow them to fluctuate naturally. Still, Eder worries some property owners will eventually demand a structure that will allow the government to manipulate water levels.

"People want the level of the lake to be six inches below the end of their dock. Well, that's not the way the system works," said Eder. "I'm damn sure people need to recognize that these are dynamic systems that need to fluctuate, and that we need to adapt."

The question under this new regime of increased precipitation and evaporation: How?

THE SPRING OF 2013 WAS REMARKABLY WET ACROSS THE GREAT Lakes, but Army Corps hydrologists said at the time that the bout of storms was only a drop in the bucket. "What we've seen this spring is what you need to get the lakes headed back," Army Corps hydrologist Keith Kompoltowicz said at the time. "You just need to see it consistently over a number of seasons."

He was wrong. Like the rest of us, he didn't see the polar vortex coming.

The following winter's arctic blast froze Lakes Michigan and Huron almost from shore to shore for the first time in decades, causing the lakes to return to their long-term average by summer 2014. Then they kept climbing. The winter of 2015 froze the lakes similarly and those two years of extraordinary ice, coupled with above-average precipitation over the same period, caused the lakes to rise more than three feet between 2013 and 2015—the biggest two-year surge on record. Then they climbed another foot the following year.

Such a dramatic rebound poses its own problems. The storms that helped restore the lakes also triggered massive sewage overflows across the region, unleashed flooding in Chicago and flushed into Lake Erie plumes of fertilizer-rich soils that spawned the record toxic algae bloom that spanned nearly 2,000 square miles.

It's becoming increasingly difficult to write off these deluges as a fluke. Meteorologists refer to the most intense rain events as "100 year storms"—tempests so severe the odds of one happening in any given year are 1 percent. Since 1997, the Milwaukee region alone has experienced six such storms, including two in just one month during

2010. "We've either had really bad luck," said meteorologist Roebber, "or something else is happening."

Roebber believes that "something else" is an era of unprecedented rains—and evaporation. He expects the average level on Michigan and Huron to continue to decline in the coming decades, but he is more concerned about how the intensified weather cycle will push the swings around the lakes' long-term average higher and lower than anyone has ever seen, well beyond the historical three-foot flux above and below their average level. Roebber said it is reasonable to expect that in coming decades the highs and lows will soar and plunge more than 4 feet from their average—meaning water levels could swing by 8 to even 10 feet.

He said a reasonable option in this increasingly unpredictable climate is to explore a gated system on the St. Clair to flush out water in wet years and hold back flows in dry years—the exact system the Great Lakes Commission's Eder fears.

Roebber acknowledges trying to shackle water levels within their historic range could be impossible. The problem is so much inertia is built into the massive lakes' hydrologic cycles that to properly operate the gates would require accurately predicting precipitation trends months ahead. If those predictions are wrong the gates may only exacerbate extreme highs and lows; water could be held back when it should have been released, or released when it should have been held back.

But Roebber said the time to explore the potential to manage the lakes in this manner is now, before disastrous levels—high, low or both—hit. "Perhaps," Roebber said, "we've reached the point where we have no choice but to fully manage the system." Pollack, the U.S. Chair of the Joint Commission, wants to do nothing of the sort. She believes it is best just to live with what nature gives and takes from the lakes. The concept in climate change circles is called "adaptive management."

"Adaptive management is a hard sell because by definition it says we don't know what it means," Pollack told me. "It doesn't start out with a prescription. It doesn't say build this and dredge that and modify the other thing." She said what it does mean is to start paying attention to the changes that have already happened, learn everything you can to better predict changes likely to come, and be strategic in your decisions on how to cope. "It is continually measuring, learning, adjusting," she said.

Some of this, of course, was already happening when the record low hit in 2013. Marinas were installing floating docks. Freighter operators, who must shed up to 270 tons of cargo for every inch of water lost in low water years to keep their hulls from cracking open on the lake bottom, were adjusting their business plans accordingly. And lakeside cities and villages were dredging their harbors ever-deeper, a type of adaptation that is costly, both in terms of dollars and environmental damage to the lakebed. I talked to the boss of one such dredging project in early 2013 who was called out to an island in northern Lake Michigan to do an "emergency" job so the ferry that is the 700-resident island's only year-round link to the outside could get into its harbor. Taking a break from the treacherous work that required him to operate a crane on a barge bobbing in an icy harbor, he talked about how similarly desperate projects were under way along Lakes Michigan's and Huron's nearly 5,500 miles of combined shoreline. He was thinking there had to be a better way—and that maybe it is to try to catch the water instead of chasing it into the boulders and bedrock.

"We can't just keep dig, dig, digging," he said.

At that moment, the extra inch Lakes Michigan and Huron gained over 100 years ago from the sinking of the *Fontana* and *John Martin* was a blessing. But by early 2016, with water levels surging well beyond their long-term average and beginning to erode the backyards of lakeshore property owners, it was beginning to look like a curse.

The story the news media took from the spike in lake levels was that things had finally returned to normal. Meteorologist Roebber said what people need to realize is that this is probably only the beginning of an entirely new "normal," one with highs and lows that the 40 million residents of the Great Lakes basin have never seen, let alone figured out how to live with.

"For the moment the lake has been restored to slightly above-average conditions," he said after the lakes surged back to their long-term average—and then continued to climb. "That's good. Nobody is going to complain. But it's not an indication that the underlying problem has been resolved."

In fact, it had only just been discovered.

Chapter 10

A GREAT LAKE REVIVAL

CHARTING A COURSE TOWARD INTEGRITY, STABILITY AND BALANCE

Ken Koyen wasn't a particularly big teenager, he wasn't a particularly tough one, and he wasn't particularly keen to carry on the backbreaking Great Lakes commercial fishing business that had sustained his family since the 1800s. But Ken's dad had other plans, especially after he asked one day to see his growing boy's hands.

"They're big enough," his dad said gruffly. "You'll do just fine."

That was almost a half century ago. Today Ken Koyen is the last full-time commercial fisherman on Wisconsin's Washington Island, which sits at the edge of a violent patch of northern Lake Michigan, known as Death's Door for the countless boats its churning currents have sunk over the past few centuries. There was a time when the island was a base for some 50 commercial fishing boats, many manned by descendants of Icelandic immigrants who found the rugged chunk of limestone where the frigid waters of Lake Michigan collide with the warmer waters of Green Bay a suitable surrogate for their North Atlantic homeland.

By 2003 Koyen was the only full-time commercial fisherman left, and the stocks that had sustained his father, and his father, and his father, and his father—lake trout, perch, chubs and whitefish—were struggling or all but gone. Koyen was left to chase what many once regarded as a trash fish, the lowly burbot—a bottom dweller that the old-timers considered nothing but a nuisance in their nets. To eat one of these fish back when Washington Island was famous for its fishing fleet was to hit bottom, a "disgrace," as one 91-year-old life-long island resident once told me. But by 2003, the islanders had realized the fresh-water cousin to ocean cod was better than nothing. In fact it was tasty enough. Delicious even. It had to be. The only other native species that drifted into Koyen's nets, which some days stretched two miles across the lake bottom, were whitefish, and they were starving, because their favored food, a quarter-inch long, shrimp-like organism that once blanketed the lake bottom, had vanished with the arrival of the invasive quagga and zebra mussels in the late 1980s and early 1990s. In the years before the mussel invasion, a seven-year-old whitefish had an average weight of nearly five pounds. By 2003 it had crashed to barely a pound. This left the famished whitefish, built to root about the lake bottom and grub what they could with their toothless mouths, with little but the sharp-shelled mussels to eat. Koyen remembers the changeover in diet as a grim one. Whitefish don't have the jaws to crack open mussel shells and suck out the meat, so they swallowed them whole and left the hard work to their stomachs, which weren't, at least at first, up to the task. Using the colorful language of a self-taught biologist, Koyen explained that the typical whitefish has an anus about the size of a "swizzle stick." But the fish excrement, a paste of crushed mussel shell thick as unset concrete, stretched the whitefish's underside orifice to the diameter of his pinky. "It actually looks like hemorrhoids," he told me. "It actually pushes part of their intestine out."

By 2005 Koyen was thinking his fishing business was similarly going down the tubes. "Honestly, I thought whitefish were done," he said, "and I thought I was done."

Then nature stepped in. Koyen said he began to notice the stomach muscle whitefish use to grind mussel shells getting bigger year after year, to the point that today he says you can see a rigid rib on the fish belly where none used to exist, and he now catches a healthy whitefish with stomachs full of mussel paste. But these native fish have done more than just adapt to a diet of mussels.

Back in the early 1980s, Koyen was fishing with his dad when the two hauled from the depths of the lake something neither man had ever encountered—a whitefish with an alewife hanging out of its mouth. What made this so strange is that whitefish are not a fish-eating fish. They don't even have teeth. "Look at that! Look at that!" Koyen's dad hollered over the rumble of the boat engine. His dad actually stopped lifting the net to get a closer look and ponder the oddity. For an old-timer who had spent his life lifting untold thousands of fish from the depths of the lake, seeing a whitefish going after another fish was about as bizarre as stumbling upon a man gnawing on a log.

Today all Koyen sees is whitefish bellies loaded with mussels and fish, particularly round gobies, a bottom-dwelling invader that might just be the future of the Great Lakes, and for one simple reason— gobies have molar-like teeth that allow them to crunch mussel shells. This means the bug-eyed fish not much larger than your thumb unlock what is otherwise a nutritional dead end for other native fish species that still cannot digest the shell—provided those native fish can, in turn, eat gobies. Koyen once cut open a whitefish that had 37 gobies in its belly and said that it is now common to hoist nets full of whitefish with their jaws ripped open from swallowing the fish whole. "They're fat," Koyen said of the improving condition of the whitefish. "They're round."

Koyen thinks he is watching evolution at work, and he isn't the only one. I talked to commercial fisherman Charlie Henriksen one sunny June morning just after he hauled in 1,600 pounds of whitefish about

30 miles south of where Koyen fishes. Henriksen said he, too, is "absolutely" convinced the species is evolving before his eyes to cope with the changes in the lake. "What we're seeing with the whitefish, well, they might be the most adaptable fish in nature," he told me. "They're more adaptable than some people I know."

The *Alicia Rae*, Milwaukee's last working commercial fishing tug, off Wisconsin's Door Peninsula.

Henriksen said the species switchover in diet from the little shrimp to mussels, gobies and any other fish they can get their mouths around means he now catches whitefish in such strange places that he has essentially had to learn to fish for them all over again, as if they were a different species altogether. He said he could not imagine describing the changes to the long-gone fishermen who taught him the trade.

"If I told them where I was fishing and what I was catching, they'd shake their heads and say: 'No way Charlie. Stop the bullshitting.' I mean, that's how much it's changed."

Biologists who manage the Lake Michigan fishery say whitefish are flourishing to the point that they are expanding their range into tributaries that feed the lake, and they now swim so thick in the open waters that an entirely new rod-and-reel recreational fishery has emerged.

"We're lucky when you consider all the crazy stuff going on, that the fish can still thrive," said Scott Hansen, a biologist with the Wisconsin Department of Natural Resources. Even crazier things are happening next door on Lake Huron.

WHEN CHARTER BOAT CAPTAINS START TELLING TALES OF THEIR most memorable catches, it's a safe bet that you can trim a few pounds off the size of the fish in the story. This was not the case with Janice Deaton. The Lake Huron charter boat captain wasn't bragging to me in summer of 2008 when she talked about landing a two-foot-long chinook salmon in the waters off Harbor Beach, Michigan. She was grieving. She was telling me about a fish that "looked more like a snake than a fish," a fish so skinny she could wrap her hand around its slimy belly. That fish was typical for the millions of chinook swimming in the lake at the time; the average weight for an adult went from over 14 pounds in the mid-1990s to barely 8 pounds by 2006. That corresponded with a drop in the lake's alewives, the favored food of chinook. And that corresponded with a 90 percent drop in the lake's phytoplankton, which, in turn, was tied to a surge in plankton-hogging quagga mussels.

This food chain reaction had left Deaton the last full-time charter boat captain hanging on in a town that only a few years earlier had about a dozen. In 2008 she was operating out of an almost vacant marina that had recently had a two-year wait list for slips. The salmon crash happened so fast that biologists have likened it to "driving off a cliff," and its impact rippled ashore. "As you're driving through town, just look side to side," tackle shop owner Art Farden said to me. "We've lost three grocery stores in the last five years." He figured Michigan's sagging economy was also to blame, but said it was no stretch to say the salmon collapse has been economically catastrophic.

"It's not natural," he said. "But it is a disaster."

Actually, it is in a way completely natural. What has happened in the decade since the crash of Lake Huron's two dominant species—invasive Atlantic alewives and the Pacific salmon planted to eat them—is an uncanny story of a Great Lake's recovery from being managed as an oversized fishing hole for nearly a half century. The chinook stocking program began in the 1960s and hummed along for decades before peaking in 2002, in terms of the number of fish caught. The next year, biologists doing netting surveys occasionally found chinook stomachs empty, a sign of declining numbers of alewives. By 2010 the chinook and alewife were all but gone. The imbalance between predator and prey destroyed both populations.

But much to biologists' amazement, the lake's native fish species surged immediately after the alewives disappeared. It turned out that alewives are death for native species in a number of ways. They gobble up the eggs and young of native species and out-compete them for the zooplankton that is the foundation of the lake's food chain. But alewives also doom lake trout in a manner that borders on subterfuge. Alewives carry high levels of an enzyme that triggers a thiamine deficiency in trout, which causes their eggs to either not hatch or induces deadly development problems in trout offspring.

The thiamine dilemma had been known for years, yet the extent that it was foiling efforts to restore native lake trout to Lake Huron with hatchery stocking was not grasped until the alewife crash. Today lake trout are again successfully breeding in the wild—so successfully, in fact, that fishery managers are considering dropping the stocking program that has kept the species on life support since the 1960s.

"It all started as soon as the alewives were gone," explained Michigan Department of Natural Resources biologist Dave Fielder. "The natives started reproducing like crazy."

The recovery of the native stocks on Lake Huron has been bolstered by their ability to feast on bottom-dwelling gobies. Just as commercial fishermen are finding gobies in whitefish bellies on Lake Michigan, it's now common on Lake Huron for sportsmen to reel in lake trout with their faces all banged up, something biologists say is a consequence of the trout rooting out gobies in the rocks on the lake bottom. Native walleye are also feasting on gobies, as are smallmouth bass, perch and brown trout, an exotic species introduced more than a century ago that is still planted today but also reproducing on its own in lake tributaries. "Those species that can make the switch to gobies are OK," said Fielder. "Those that couldn't, they're gone. And the chinook couldn't."

The remarkable result is that the top of the Lake Huron food chain more closely resembles its natural self today than any time since the lamprey and alewives invaded in the mid-1900s. It has everything to do with the disappearance of alewives, and little to do with Howard Tanner's and Wayne Tody's grand salmon plan. Tanner always said he never intended to use the salmon to eliminate the alewives. He saw them, in fact, as the foundation of the lakes' future.

"You wanted the alewife to be alive and healthy. You wanted them there forever," said Tanner. "Maybe not as such a big nuisance, but that was the foundation of what you were trying to build. We didn't want to destroy it."

Some biologists today are looking at alewives completely differently. One of them is Fielder, who has on his cluttered desk a "Tanner and Tody Award" he received from the state of Michigan for examining what's happened to Lake Huron's food chain since the alewife collapse. His conclusion: "If you really want a native fish recovery, you're not going to fully achieve that in the presence of alewives, unless natives are sustained by hatcheries. And how can you call that a recovered fishery?"

The return of the natives has not occurred without economic pain; the top fishing towns along Michigan's Lake Huron coast have lost, collectively, a minimum of $19 million per year since the chinook collapse. It is reflected in empty marinas and hotels, closed bait shops and lonely harbors. Yet some Lake Huron communities and businesses have begun to adapt, right along with the fish.

Ernie Plant's eyes got wide when I walked into Frank's Great Outdoors gear and bait shop north of Bay City in fall 2014 and I asked him what the salmon fishing on Lake Huron was like during its peak in the 1980s. The salmon, he explained, tied him not just to the lake, but to his father, who would take him on overnight fishing excursions after his Friday night high school football games. They would spend fall weekends along the shore chasing the chinook that were chasing the alewives, which ran so thick they clogged water-filled ditches along coastal roadways. "We never had a boat," said Plant, now a sales manager at the outdoor shop. "But we didn't need one."

Those salmon fishing trips with his dad helped inspire Plant to pursue a degree in biology from Northern Michigan University, but when I asked him whether he'd welcome a return of the salmon of his memories, he paused and looked up at the ceiling of the sprawling store bursting on an early fall Tuesday morning with fishing poles, electronic fish finders—and customers. "I don't know," he said after taking a deep breath. "We've adapted."

Chinook might have been a boon for charter captains with giant boats that could handle the ornery lake's open waters, Plant said, but regular people can fish on their own for walleye. "A lot more people fish for walleye than chinook," he explained. "It's more economical. The equipment costs less and the boats don't have to be so big because you don't have to go out so far."

The big question now is if the same salmon–alewife collapse and subsequent surge in native species will happen next door on Lake Michigan. The collapse side of the equation is already under way, leading many to wonder if Tanner's salmon playbook, which worked so well for so long, will even be relevant in the coming decades. "The salmon are from the Pacific. The alewives are from the Atlantic. You put them both in the Great Lakes," United States Geological Survey Chuck Madenjian told me in 2015 after an annual Lake Michigan prey fish survey turned up almost no alewives. "Who's to say that's a situation that is going to last forever?"

Nobody anymore.

THE PROSPECTS FOR A GREAT LAKES–WIDE NATIVE SPECIES renaissance similar to what has happened on Lake Huron may prove fleeting. This is because the doors to new Great Lakes invasions—and the fresh ecological chaos they trigger—remain stubbornly open more than a quarter century after overseas freighters sailing up the St. Lawrence Seaway unleashed the zebra and quagga mussel infestation. In 2013 the Environmental Protection Agency, more than 40 years after Congress passed the Clean Water Act, finally agreed to follow that law and require ships visiting the lakes and other U.S. waters to install ballast water treatment systems. But few who understand the challenges of sterilizing ballast tanks, including the three federal judges who in 2015 unanimously ordered the EPA to come up with stiffer treatment standards, believed the agency's pollution rules for ballast water were nearly stringent enough. The court, however, gave the EPA no deadline to issue the new rules, and, in the meantime, the judges let stand the existing treatment requirements, which don't demand treatment systems for all ships visiting the lakes until 2021, at the earliest.

The shipping industry likes to point to the fact that no new Great

Lakes invaders have been discovered since 2006, before overseas ships sailing up the Seaway were required to flush their ballast tanks in mid-ocean to expel or kill any freshwater hitchhikers. This is at best a misguided argument given the history—and cost—of ballast-borne invaders.

This "front door" to Great Lakes invasions could be shut without a single new technological development. In 2015, only 455 overseas ships sailed up the Seaway, an average of a little less than two overseas ship per day. Force those ships to transfer their cargo to local ships or railroad lines (which insist they have more than ample capacity to handle the business) without entering the Seaway, and the door is effectively shut. The economic equation is clear. The cost of future invaders, which we know from painful experience, could be in the billions of dollars—and who knows at what ecological cost. The additional cost to shippers pales in comparison and could be compensated for in any number of ways.

Then there is the "back door"—the Chicago Sanitary and Ship Canal, the prime pathway for two species of Asian carp to make the jump from the carp-infested waters of the Mississippi River basin to the Great Lakes. The federal government spent more than $318 million between 2009 and summer 2015 to block the carp's advance into the lakes. That included some $25 million on a 10,000-page study on how to rebuild the natural divide between the Great Lakes and Mississippi River basin that the Chicago canal obliterated when it opened in 1900. The 2014 study prepared by the U.S. Army Corps of Engineers—an agency in the business of moving barges and not controlling unwanted fish—was in many ways a farce. The study concluded that plugging the canal system would cost up to $18 billion and take decades to complete. Yet the bulk of those dollars were yoked to projects that have little to do directly with stopping the fish. Some $12 billion of it was set aside to build new reservoirs, sewer tunnels and water treatment plants as well as remove contaminated river sediments. These are all worthy projects but they are not directly tied to stopping the fish. Critics saw them as a cynical ploy to make the project as unappealingly expensive as possible.

"It could be read," said Natural Resources Defense Council's Henry Henderson, the former commissioner on environment for the City of Chicago, "as a laundry list of why this can't be done." Henderson, along with groups representing the region's mayors and governors, contends the project could be done for a fraction of the Army Corps' estimate, but nobody is expecting any earth to be turned anytime soon.

And although biologists debated for years the true damage the two plankton-feasting Asian carp species—bighead and silver—could do to plankton populations in the already-ravaged lakes, a study released in late 2015 concluded the fish could dominate Lake Erie to the point that for every three pounds of fish in the lake, one pound would be Asian carp. At the same time, there is a third species of Asian carp that hasn't yet made headlines but is also slowly making its way toward the Great Lakes. Black carp were imported by federal officials more than two decades ago to help Southern fish farmers keep their catfish ponds snail-free. These fish aren't like the other two species of Asian carp, which filter plankton out of the water. They can grow to a bear-sized 150 pounds and do so by eating mollusks—including zebra mussels.

Canadian politicians have supported plugging the Chicago canal to stop the carp but have shown no similar interest in getting equally tough with the Seaway, which it jointly owns with the United States. This means the U.S. government is the only authority with keys to both the front and back doors, and even though Congress has shown little desire to shut them anytime soon, it spent some $2 billion between 2010 and 2015 for an ongoing Florida Everglades-style "Great Lakes Restoration Initiative." The program is focused on cleaning up toxic sediments, restoring wetlands, stopping the flow of toxic algae-fueling agricultural nutrients into the lakes and controlling unwanted species. The money and the programs—including some strange ones, like the more than $5 million set aside to trap and shoot feral pigs in Michigan—are mostly welcomed and helpful, but the Great Lakes need more than money.

The program sometimes brims with painful irony:

- Even as the EPA spends billions of dollars on restoration programs, including efforts to combat invasive species, conservation groups have had to take the agency to court to force it to enact adequate ballast treatment regulations to prevent new invasions.
- Even as the U.S. Fish and Wildlife Service touts a "zero tolerance" policy for new invasions and assists the EPA in administering the restoration initiative, it too has been in court, arguing that federal efforts to stop the carp are adequate at the same time the genetic evidence mounts that at least some fish have breached the electric barrier on the Chicago canal.
- Even as politicians across the Great Lakes region talk about the need to stanch the flow of algae-fueling nutrients from croplands into the lakes, they refused to amend the Clean Water Act to regulate "non-point" pollution. Congress's own Government Accountability Office released a report in 2014 that pointed squarely at the exemption and declared that until lawmakers fix this loophole, the nation's waters will remain woefully polluted. "There is a deficiency of governance here," said Henderson, "that goes beyond funding."

Here is another way to look at it: Before you start administering chemotherapy to a lung cancer patient who is a lifetime smoker, wouldn't it be wise to first try to get the patient to stop smoking? There are indeed new medicines in the works that could undo the damage wrought by the biological pollution of contaminated ballast water, ones so powerful they could make the blind and desperate—and ultimately successful— hunt for a lamprey-specific poison look like mere ecological tinkering.

THE BIOLOGICAL BOMBSHELL THAT ARRIVED AT AN AUBURN UNI-
versity laboratory in 2009 came in the tiniest of packages—a cardboard
box about the size of a paperback book. Inside it were several plastic
tubes that contained flecks so small that they were invisible. Lab workers tipped these tubes into a solution that was as innocuous looking as
a glass of water. It was anything but. In it floated a designer "poison"
potent enough to eradicate an entire fish population.

The concoction, which held a specially modified fish gene built in
a high-security laboratory on the Australian island of Tasmania, was
placed into Petri dishes thick with *E. coli* bacteria and then dosed with
a chemical that allowed the bacteria to absorb the genetic material.
As the fast-reproducing *E. coli* numbers then exploded, copies of the
gene replicated right along with it, with two new genes emerging each
time a bacteria cell split. In a matter of hours the Auburn biologists
had untold millions of *E. coli*, each carrying a strand of the manmade
genetic code. And each of those strands held the power to take a species' collective sex drive and throw it in reverse, to turn sexual reproduction from a life-sparking act into a life-snuffing one.

It does so by adding a twist in the DNA so a fish implanted with the
gene can produce only male offspring. The concept, called the "daughterless gene," is devilishly clever: a developing carp turns female only
after an enzyme transforms the male hormone androgen into the
female hormone estrogen. This gene blocks production of that enzyme,
so the embryonic fish cannot make the early-life transformation from
male to female. The idea is that if you plant enough of these daughterless fish into a lake or river for a sustained period, it's just a matter
of time until it runs out of females to carry on a population. The fish
breed themselves to oblivion.

It is no surprise this anti-carp technology was pioneered in Australia, a continent ravaged by invasive common carp that were imported by fish farmers. Those fish eventually—as almost always goes the story—escaped their containment ponds in floodwaters. It happened in the 1970s, right around the time Asian carp got loose in the United States. And as is the case with Asian carp in some stretches of river in the Mississippi basin—North America's largest—common carp now account for as many as 90 percent of the fish in some areas of the Murray-Darling basin, Australia's largest river system.

Ron Thresher, an American who pioneered the DNA-based pest control at a research center in Tasmania, Australia, explained to me on a tour of his lab that the genes he works with are species-specific and pose no measurable risk to the public or any other type of fish or organism. Still, the lab's security was extraordinary. To prevent any of the specimens carrying the gene from escaping into the wild, the lab windows were sledgehammer-proof. The floors near the doors were raised in a manner so no water—or fish—could float out if a tank burst. The walls were waterproof to a level that they could contain all the water in all the aquariums holding the genetically manipulated fish. Pipes from the lab didn't drain to the local sewer system. They led to a boiler for sterilization. If that boiler failed, there was a twin backup.

An obvious worry is that if just one of these fish implanted with the daughterless gene were set loose in the wild the gene could, eventually, spread across the globe and ultimately eradicate all females of the targeted species in a manner that would doom it to extinction. But Thresher said his team specifically designed their carp gene so that once it reaches the third generation, the fish that carry it only automatically generate males in half of their offspring. The other half of their progeny is half male and half female. The gene's impact would be halved similarly in the next generation, and so on, all in a manner that its effect on any fish population's sex ratio would disappear over time. This, Thresher explained, means hatchery-raised stocks of the daughterless fish would have to be planted for multiple generations to knock out a local population.

(Thresher said it is in fact possible to design a gene that ensures all-male progeny in every subsequent generation in a manner that could spark an extinction. "In theory, the release of a single carrier of one of these could doom a population. Not surprisingly, there is considerable debate about the risk," he told me. "We deliberately went for our approach because it is inherently much safer and we thought it wise to walk before one runs, in terms of recombinant genes to manage an invasive pest.")

After Thresher proved the gene worked in a type of fast-reproducing minnow in his lab, the next step was to figure out how to do the same thing with common carp, a much more daunting project given the larger size of the fish and the fact that it can take them years to reach sexual maturity. This means that tracking the gene down through several generations requires the better part of a decade, or more. A carp experiment similar to the one done on minnows in Thresher's lab also had to be scaled up from aquariums to ponds to see if the genetically modified fish could woo normal females in a more natural setting; the gene is useless, after all, if their carriers fail to get it into the next generation.

Thresher found a partner in Auburn's Rex Dunham, who in 2009 cleared the paperwork with his university to take delivery of Thresher's carp-eliminating gene. By early 2015 he was preparing to produce a third generation of daughterless carp at a gated research facility just north of the Auburn campus in eastern Alabama, about two hours south of Atlanta. The largest, most sexually matured fish—all males— were already swimming inside a fenced-in fish pond inside a barbed wire-capped chain-link fence that pulsed with low-voltage electricity to keep out potential predators like raccoons. The pond was also further covered with fine black netting to keep birds from plucking a fish from the water and lifting it off alive. The pond drained through a pipe with a screen on it. That pipe drained into a lower pond that drained through a sieve too fine for a large fish to flush through. That pond, in turn, flowed into a reservoir that had been purposely stocked with predator fish in case fish from ponds and tanks above somehow made it that far. Dunham called the whole setup "zealous," but it was clear he, like Thresher, takes the security issue seriously.

Dunham received about $100,000 from the Australians to conduct early research with his carp, but by the time I visited all that funding had dried up and he was doing the work basically for free. He said he was motivated by the mess invasive species have made across the globe—common carp in Australia; pythons in Florida; Asian carp in the rivers of his home state of Illinois.

"To have the potential to rectify our past environmental mistakes, to me, it's just intriguing and exciting that we could go back and do that," he said. "There are very few proven ways to go back and fix your mistakes, and the potential here is to eradicate a nuisance fish species, without poisoning a river or a lake or harming other fish. To me, that's an ethical thing to do."

Thresher was quick to acknowledge the potential for political backlash for planting daughterless fish in the wild and said he is happy to leave it up to the politicians to decide whether, when, where and how to deploy such a genetic tool. It's a decision that may take center stage sooner than people expect. Thresher predicts genetic solutions to invasive species problems would be "widely available" by the early 2020s, and he wasn't talking just carp. "The basic technology, once it's up and running," he said, "I think will be applicable to a wide range of things."

Mussels included?

"Mussels included."

This has Russ Van Herick intrigued, and not a little wary. He heads a think tank funded by the Great Lakes states to develop creative solutions to some of the lakes' most vexing problems, and that has him thinking about how to use—or not use—a DNA-based eradication tool that he, too, is convinced is on the horizon. He wonders who will make the decision to deploy it. In the 1960s the state of Michigan decided on its own to plant exotic salmon, even though it was clear at the time the fish would not stay contained in Michigan waters. Would a single Great Lakes state today try to act on its own and release a manmade gene in a similar fashion? If not, would it take a unanimous vote by all the Great Lakes states? What about the Canadian provinces? What about the federal governments? What about the prospect of mischievous, if well meaning, individuals or groups acting on their own?

"We are not even close to developing a governance system to catch up with these emerging technologies," he said.

And if science does someday allow us to recast the characters in the Great Lakes in a manner never before possible, how do we decide what those characters should be? Do we continue to manage the lakes for maximum sport fishing fun and maybe even modify certain species so they fight harder and have tastier flesh? Do we farm the lakes for energy-producing, genetically modified algae? Or do we try to resuscitate any and all native species in any way possible?

One person long ago figured out where to start.

"A thing is right when it tends to promote the integrity, beauty and stability of the biotic community," famed Wisconsin naturalist Aldo Leopold wrote in 1949, which happened to be the peak of the lamprey invasion. "It is wrong when it tends otherwise."

Perhaps that's as good a place as any to start.

Finding these "right" things may take no genetic tinkering at all. It may just require that we let the lakes heal on their own by protecting them from fresh invasions and allowing the fish, mussels and microorganisms that are already here find a new ecological balance. It is, after all, already happening, despite the salmon advocates' best efforts to keep their artificial fishery humming.

HOWARD TANNER WAS NEVER BIG ON THE IDEA OF VALUING NATIVE species simply because they are native. His priority in the 1960s was to convert the lakes from a resource primarily managed as a commercial fishery into a sportsmen's haven, and native species just didn't fit his bill—and they still don't.

"I doubt if the charter boat captains can sustain a fishery on lake trout," he said.

He called trolling for native walleye "about the most boring thing you can do."

"Like bringing in a wet sock," he said.

This fun-first mentality has worn thin on Tom Matych. He isn't a fisheries biologist. He doesn't even have a college degree. But the retired construction worker from western Michigan thinks big, like Tanner. And he has a plan for a Great Lakes recovery that does not require manipulating fish genes or the perpetual stocking of exotic fish. He wants lake managers to change their focus from a system dominated by chinook to one that can sustain as many native predators as possible. He's not just thinking about more species for fishermen to catch; he's thinking about the ecological health of the lakes. Matych considers native predators to be the lakes' "biotic resistance" to new invasions, noting that species like lake trout, perch, walleye and smallmouth bass all feed on different things at different times in different places. Big numbers of these fish, he maintains, make the lake more resistant to new invaders, including Asian carp. He'd like to start with a program to stock the shoreline areas with native perch.

"We want to do the same thing that Tanner did," said Matych. "But with native predators for the 180-something invasive species that the salmon don't eat." That 180-plus figure for invasive species in the Great Lakes is actually a tally of all the lakes' nonnative organisms, including the salmon that have been stocked. But the gist of his argument is not lost on a swelling number of Great Lakes biologists.

"There is a body of evidence that says having a robust, co-evolved food web helps prevent invasive species' establishment," said John Dettmers, a biologist with the Great Lakes Fishery Commission, which helps Great Lakes states and provinces coordinate fisheries management. "It doesn't guarantee it, but it's something to be considered by managers if they are concerned about invasive species."

Matych doesn't think the end of salmon means the end of Great Lakes fishing. He remembers how, as a child in the 1960s, his nose could tell him when the native perch were running especially thick on the eastern shore of Lake Michigan near Muskegon at certain times of year.

"If you could smell the paper mill, which smelled like rotten eggs, you knew the perch were coming, because the west wind blew in the zooplankton and the minnows, and the perch followed them in," he said. "I mean, people took those days off work." Matych remembers how people rode the public bus with bamboo fishing poles, and how that bus would stop at a bait shop near the lakeshore and how crowded the pier was and how he felt "10 feet tall" as a five-year-old when he caught his first perch with his father.

It's an experience he wants to share with his own granddaughter, but Lake Michigan's perch fishery collapsed in the 1990s. Michigan and Wisconsin state fishery biologists blame a complex set of factors, including a loss of perch-sustaining plankton tied to the quagga mussel invasion. Matych points a finger straight at the states' decades-old efforts to produce a maximum number of chinook, which has required biologists to essentially manage alewives as a protected species rather

than the troublesome invader it is. When Lake Michigan alewife numbers dipped in the 1990s, for example, the state of Wisconsin banned
commercial harvest of the fish, which were being sold for cat food and
fertilizer. Alewife numbers rebounded; the lake's native perch population crashed.

"I'm not good at being politically correct," he told me. "But when
they say there is not enough food for perch, what they're really saying is there is not enough food for perch and alewives." He cited the
energy—the number of little fish—it takes to make one big chinook.
"You need 123 pounds of alewives to make a 17-pound chinook, and
you need 40 pounds of zooplankton to make one pound of alewives," he
said. "That's a whole lot of zooplankton for one fish." (I checked these
numbers with fisheries biologists, and they check out.)

It's no surprise the states that manage the Lake Michigan fishery
maintain their focus on the species sportsmen want; in Wisconsin
alone, nearly $23 million of the state's $25 million fisheries budget for
2014 came from fishing licenses and taxes on fishing-related products.

The federal government, not similarly financially tied to the sport
fishing industry, has been stocking up to three million lake trout annually in Lake Michigan as part of a Great Lakes native species recovery
program that dates to the early days of the lamprey poisoning in the
1960s. The Lake Michigan lake trout restoration effort had dragged on
ever since with little sign the fish were reproducing on their own—
until the lake's alewife population started to plummet. Several years
ago small numbers of lake trout were found on a reef in the middle
of Lake Michigan without the telltale fin clip that is the sign of a
hatchery-raised fish. By 2015 federal biologists were finding as many as
50 percent of the lake trout they netted in some areas of Lake Michigan
were naturally reproducing.

With evidence piling up that what is good for native lake trout (no alewives) is bad for salmon (which are alewife-dependent), and vice versa, the states and federal government are trying to strike a dicey balance to keep sportsmen happy and lake trout on the road to recovery.

"The idea is that you keep enough alewives so you have a chinook fishery but not so many that you don't have natural lake trout reproduction. Well, that's a razor-thin line," said Dale Hanson, a biologist with the U.S. Fish and Wildlife Service working to restore Lake Michigan's lake trout. "I don't even know if it's possible."

It wasn't on Lake Huron.

University of Wisconsin–Milwaukee fisheries biologist John Janssen calls Tanner's idea to convert the alewife infestation into a popular and profitable salmon fishery a "brilliant" decision—for its time. Now, he said, it's time to view the invasive gobies in a similar way—as sustenance for the fish that can feast upon them, which is mostly native species, as well as exotic brown trout. None of these fish put up a fight like a hooked salmon, but to disparage the very species that are adapting to the mussel-altered lakes, Janssen said, is to risk losing the next generation of anglers. He told me about the blowback he got from fishing guides at a meeting not long ago to discuss restoring walleye to the western shore of Lake Michigan. The guides fought it because they were worried it would further dent their struggling chinook business.

"The problem," the 60-something Janssen told the guides, "is you guys are all old. I'm old. We're all going to be dead pretty soon. The big issue is the 12-year-old on the bike. What's he going to be catching?"

That 12-year-old, you see, is perhaps the best hope the lakes have to recover from two centuries of over-fishing, over-polluting and over-prioritizing navigation: almost every person I've ever talked to who cares anything about the lakes and the rivers that feed them does so because they have a childhood story about catching the fish that swim in them.

A generation ago, Brian Settele was that 12-year-old, rolling his Schwinn 10-speed down a bike trail converted from an old railroad bed through the Milwaukee suburbs to the city's harbor with fishing rods lashed to the handlebars. His parents were going through a divorce and he found sanctuary along the urban waterfront reeling in perch and the odd salmon that drifted into the harbor.

Today Settele is a licensed captain running charters out of Milwaukee. He'll chase chinook if that's what the customer wants. But mostly Settele has learned to make a living by focusing his business on brown and lake trout—two species thriving in the evolving environment. These trout can't bend a rod like chinook—nothing does, Settele says—but they have figured out how to have a future in the lakes, and they still keep the clients rolling in.

One customer in fall 2014 was my son, first-time Lake Michigan fisherman John Egan. Age 8. The water that pre-dawn October morning was so black it was invisible as Settele steered his boat through the north gap of the Milwaukee harbor break wall into the early morning swells. Not long after a flaming orange sun popped on the horizon, one of his reels whirred and Settele yelled for somebody to grab the bouncing rod. John scurried up and snatched it. He felt the line tug somewhere deep below what had transformed from black swells to choppy gray-green waves and began to crank the reel.

You could see in the wrinkles on his nose and in his darting eyes that he wasn't struggling so much as soaring. It was both grimace and grin—an expression he doesn't have when playing computer games or chasing a soccer ball. After huffing and puffing and cranking the reel for a couple of minutes, his eyes popped and jaw slackened as a freckled three-pound brown trout broke the surface. It might have been a scrawny thing compared to one of Tanner's Pacific beasts.

But it was John's fish, caught on John's lake.

And the hook was set.

The author's son, John Egan, on Lake Michigan with his first catch.

ACKNOWLEDGMENTS

Properly acknowledging all the guidance and encouragement I received while researching and writing this book would take another book because these pages are essentially the sum of more than a decade of covering the Great Lakes full-time for the *Milwaukee Journal Sentinel*. But tops on the list is *Journal Sentinel* editor George Stanley, who first gave me a job, then pushed me to turn my newspaper work into a book, and then gave me the time off to do just that.

Columbia University professors Sam Freedman and Jonathan Weiner got George's book idea rolling; Sam taught me how to put together a book proposal; Jonathan taught me how to write my first book chapter. Agent extraordinaire Barney Karpfinger refined that proposal and wisely steered it onto the desk of Matt Weiland, who pushed (and pushed) me to elevate the writing and structure of the book for a wider audience, one much less familiar with the lakes than are readers of the *Milwaukee Journal Sentinel*. I gave Matt 400 messy pages in June 2015. More than a year later, and countless thumps on my doorstep from UPS envelopes bursting with his marked-up pages, I gave him this book.

Institutionally, I'd like to thank Marquette University for granting me an O'Brien Fellowship in 2013–2014 that allowed me to write three newspaper series that later became the foundation for three chapters. Columbia University Graduate School of Journalism gave me a Robert Wood Johnson Fellowship in 2011–2012 that allowed me to bring my family to New York for a year for what turned out to be a grueling but rewarding mid-career "break." Several months into my book-writing leave, Columbia and Harvard Universities gave me a $30,000 Lukas Work-in-Progress Award in early 2015. It was money that could not have come at a better time, a time when I was drawing no paycheck from the newspaper and my four young children would not stop eating and their feet would not stop growing.

Val Klump at the University of Wisconsin–Milwaukee invited me to campus as a visiting scholar during my 18-month leave from the newspaper, which meant an office and library privileges. St. Lawrence University's Mark McMurray and Paul Haggett made me feel at home in their archives while I researched the history of the St. Lawrence. The good people at the Carey Institute for Global Good in Rensselaerville, New York, housed and fed me (exquisitely) in early winter 2016 so I could do nothing but write.

My parents, Dick and Anne Egan, let me use their often-vacant apartment (with an inspiring view of Lake Michigan) in downtown Milwaukee to write. They also read countless early chapter drafts. Rex Dunham, Rob Gess, Noah Hall, John Janssen, Dan Macfarlane, Don Scavia, Gerald Smith and Randy Schaetzl were early readers of some of the more technical material, and their deep knowledge and patience helped me to interpret complex subject matter for lay readers; any errors in these translations are my own. *Journal Sentinel* colleague Meg Kissinger was a constant source of encouragement. Authors Tom Zoellner and Tim Weiner were most generous in offering wisdom for getting through the peculiar torture that is turning 400 pages into a book. And thanks to my wife, Alice, who osmotically suffered that same torture, but with much more equanimity.

NOTES

This book is the sum of more than a decade of reporting for the *Milwaukee Journal Sentinel*, where I have covered the Great Lakes as my full-time beat since 2003. The job has taken me across the Great Lakes and thousands of miles beyond their shores—to places as far-flung as the red rock desert of Utah, the wilds of Tasmania and the western cliffs of Ireland.

In my early years covering the lakes, I was not reporting and writing with an eye on someday turning the material into a book. That changed in 2011 after a fellowship at Columbia University during which I enrolled in a book-writing seminar that required preparing a book proposal. I put the material covered in several major series I had written for the newspaper into what was, for me, a logical order, a rough form of a book. It worked, so thought my book seminar professor, Sam Freedman, my eventual agent, Barney Karpfinger, and my book editor, Matt Weiland. The newspaper reporting and book writing blended after that, particularly during 2013 and 2014 when I was an O'Brien fellow at Marquette University doing a major project on Great Lakes invasive species that would, eventually, appear in the *Journal Sentinel*. I was encouraged to take the Marquette fellowship by my newspaper editor George Stanley, who told me at the time—"now you can start writing chapters for the book." So, many of the interviews, and some of the material within some of the chapters, originally appeared in the pages of the *Journal Sentinel*. Putting all this into book form—and one written for a national audience and not a Great Lakes–familiar regional one—required much additional reporting, writing and organization done between December 2014 and June 2016.

INTRODUCTION

xi **In 1634 voyageur Jean Nicolet**: C. W. Butterfield, *History of the Discovery of the Northwest by Jean Nicolet in 1634 with a Sketch of His Life* (Cincinnati: Robert Clarke & Co., 1881), 59–60.

xi **in a flowing Chinese robe**: There is debate about whether Nicolet actually believed he had paddled to China. The whole trip, after all, was over freshwater. In explaining Nicolet's rationale for donning the robe, Butterfield, in his *History of the Discovery of the Northwest*, noted it was because "possibly, a party of mandarins would soon greet him and welcome him to Cathay. And this robe—this dress of ceremony—was brought all the way from Quebec, doubtless, with a view to such contingency."

xii **The biggest lake in France**: Lac Du Bourget is the largest lake in France—in summer.

xii **when Northwest Airlines flight 2501**: U.S. Department of Transportation Civil Aeronautics Board Accident Investigation Report, January 18, 1951; *Milwaukee Journal Sentinel*, March 10, 2014.

xiii **Roughly 97 percent of the globe's water**: These oft-cited numbers are very rough estimates of the global water budget. Although the five interconnected Great Lakes are, by far, the largest expanse of water in terms of surface area, Siberia's remote Lake Baikal, because of its depth, holds approximately as much water as all the Great Lakes combined.

xiii **"The wars of this century have been fought over oil"**: Serageldin recounts his 1995 speech in Stockholm in the journal *Nature*, May 14, 2009.

xv **no statutory authority from Congress to make this exemption**: Author interview with Nina Bell, executive director for the Northwest Environmental Advocates; Daniel E. O'Toole, *William and Mary Environmental Law and Pol-*

icy Review 19, no. 1, 1994, 12–13; for a deeper discussion of the ballast exemption, see Jeff Alexander, *Pandora's Locks: The Opening of the Great Lakes—St. Lawrence Seaway* (East Lansing: Michigan State University Press, 2009), 84–87.

xvi **"These ships are like syringes"**: *Milwaukee Journal Sentinel*, October 31, 2005.

xix **What in the hell were they thinking?**: Tim Heffernan, Bloombergview, August 31, 2012; The *Hutchinson* (Kansas) *News*, July 5, 1907.

CHAPTER 1. CARVING A FOURH SEACOAST

3 **legendary CBS newsman Walter Cronkite**: *The Eighth Sea*, a 30-minute promotional color film sponsored by Caterpillar Tractor Co. in cooperation with the U.S. and Canadian Seaway authorities, Ontario Hydro, the Power Authority of the State of New York and the U.S. Army Corps of Engineers.

4 **Some six million years ago**: Daniel Garcia-Castellanos et al., "Catastrophic Flood of the Mediterranean after the Messinian Salinity Crisis," *Nature*, December 2009, 778–781; author email interviews with Garcia-Castellanos.

6 **About 7,600 years ago**: William B. F. Ryan and Walter C. Pitman, *The Black Sea: Noah's Flood: The New Science Discoveries about the Event that Changed History* (New York: Simon and Schuster, 1998).

11 **The ditch-digging**: Ronald Stagg, *The Golden Dream: A History of the St. Lawrence Seaway* (Toronto: Dundurn Press, 2010). There are several excellent books I used to inform me on the history of the St. Lawrence River-turned-Seaway, including Daniel MacFarlane's *Negotiating a River: Canada, the U.S. and the Creation of the St. Lawrence Seaway* (Vancouver: UBC Press, 2014) and Carlton Mabee's *The Seaway Story* (New York: Macmillan, 1961). A trove of historical documents on the construction of the Seaway is the Mabee archives housed at St. Lawrence University in Upstate New York.

14 **"I need not remark to you Sir"**: George Washington's letter to Virginia Governor Benjamin Harrison, October 10, 1784. *The Writings of George Washington from the Original Manuscript Sources, 1745–1799*, Vol. 27 (United States Government Printing Offices, 1934), 475.

15 **Jesse Hawley, a flour merchant**: Peter L. Bernstein, *Wedding of the Waters: The Erie Canal and the Making of a Great Nation* (New York: W. W. Norton, 2005); David Hosack, *Memoir of De Witt Clinton* (New York: J. Seymour, 1829), 301–341.

16 **On October 26, 1825, at precisely 10 a.m.**: Martha Joanna Lamb, *History of the City of New York, its Origin, Rise and Progress* (New York: A. S. Barnes, 1877), 696–702.

17 **far more portentous:** Buffalo *Journal*, Nov. 29, 1825 (reproduced in Vol. 14 of the Buffalo Historical Society Publications, 1910), 388.

18 **In its first year alone there were about 7,000 boats:** Bernstein, 325.

18 **By 1845 more than 1 million tons:** Ronald E. Shaw, *Canals for a Nation and the Canal Era in the United States 1790–1860* (Lexington: The University Press of Kentucky, 1990), 46.

21 **"Nature has already done most of the work":** Hanford MacNider, former U.S. ambassador to Canada, *The Rotarian*, March 1939, 21.

21 **"Our good neighbors to the south":** *Canadian Broadcast Corporation*, July 26, 1951.

21 **"If Canada proceeds unilaterally":** *A Report to the National Security Council by the NSC Planning Board on National Security Interests in the St. Lawrence–Great Lakes Seaway Project*, April 16, 1953. Dwight D. Eisenhower Presidential Library, Museum and Boyhood Home, Abilene, Kansas.

22 **with a pen that held a piece of timber from old Fort Detroit:** *Schenectady* (New York) *Gazette*, May 14, 1954.

22 **the "Gentleman":** *The St. Lawrence Seaway Collection at St. Lawrence University, Carleton Mabee Series*, St. Lawrence Seaway Development Corporation press release, June 9, 1955.

23 **In the summer of 1955:** *The St. Lawrence Seaway Collection at St. Lawrence University, Carleton Mabee Series, Newsweek*, August 15, 1955.

23 **The prospects of the Seaway:** *The St. Lawrence Seaway Collection at St. Lawrence University, Carleton Mabee Series, Time*, June 6, 1955.

23 **"the greatest single development":** *Milwaukee Journal Sentinel*, October 30, 2005.

23 **In Detroit, Chrysler was predicting:** Associated Press, as appeared in the *Corpus Christi Caller-Times*, December 25, 1955.

23 **"The Seaway will pull Europe Closer to Duluth":** *Winona* (Minnesota) *Republican-Herald*, May 7, 1954.

24 **"The only thing that made New York the biggest city":** *United Press*, as appeared in the *Victoria Advocate*, December 22, 1954.

24 **a much more dubious distinction:** Daniel McConville, *Invention and Technology*, fall 1995, as appeared in the *Milwaukee Journal Sentinel*, October 30, 2005.

25 **"The majority of general cargo seagoing ships":** *The St. Lawrence Seaway Collection at St. Lawrence University, Carleton Mabee Series*, St. Lawrence Seaway Development Corporation press release, November 15, 1954.

26 **There has to be a better way:** Brian Cudahy, *The Container Revolution—Malcolm McLean's 1956 Innovation Goes Global*, Transportation Research Board of the National Academies NEWS, September–October 2006.

28 There were three-day-long jams: *Reading* (Pennsylvania) *Eagle*, May 31, 1959.

29 "The St. Lawrence Seaway, dream of": Ibid.

29 "disgustingly small number": *Milwaukee Journal Sentinel*, October 30, 2005.

29 "The Seaway—I like to forget it": Ottawa *Journal*, June 16, 1970.

30 By 1982 Seaway revenues were so sluggish: *Chicago Tribune*, December 21, 1982.

30 the Seaway could only: *Milwaukee Journal Sentinel*, October 30, 2005.

30 "The overriding, overwhelming regret is that we built it too small": Ibid.

31 But the overseas component of the Seaway's traffic: John C. Taylor, "The Cost-Benefits of Ocean Vessel Shipping in the Great Lakes: Value to Industry vs. Environmental Damage," *Seidman Business Review* 12, no. 1, 2006.

31 "Upon arrival in a Southern California port": Jon S. Helmick's testimony before the U.S. House of Representatives' Subcommittee on Coast Guard and Maritime Transportation Committee on Transportation and Infrastructure, May 3, 2000.

32 "Assuming that compressed air were pumped": *New Scientist Magazine*, April 3, 1958, 12.

32 The concept was scrapped: James L. Wuebben, ed., *Winter Navigation on the Great Lakes: A Review of Environmental Studies*, U.S. Army Corps of Engineers, May 1995.

32 "There is not a great deal we can do about it": *Watertown Daily Times*, April 24, 1975.

33 "the world's most expensive seagull roost": *Milwaukee Journal Sentinel*, October 30, 2005.

33 If you drive down Ontario's King's Highway: I traveled to Massena, New York, and Cornwall, Ontario, to visit the Lost Villages in summer 2005 and again in spring 2015.

34 "'We'll be moving tomorrow'": *Milwaukee Journal Sentinel*, October 30, 2005.

34 The new lake swallowed about 38,000 acres: Murray Clamen and Daniel Macfarlane, "The International Joint Commission, Water Levels, and Transboundary Governance in the Great Lakes," *Review of Policy Research* 32, no. 1, 2015.

35 eighth-grade student Pat Kenney: *The St. Lawrence Seaway Collection at St. Lawrence University, Carleton Mabee Series*, St. Lawrence Seaway Development Corporation press release, June 8, 1955.

CHAPTER 2. THREE FISH

37 **There was a time when a river flowed**: Shelley Dawicki, *Oceanus Magazine* 44, no. 1, June 10, 2005.

38 **At the base of the food web**: Special thanks to fisheries biologists John Janssen of the University of Wisconsin–Milwaukee, Gerald R. Smith of the University of Michigan and Henry Regier, professor emeritus of the University of Toronto, all of whom, in interviews and emails, patiently walked me through the history of the early lakes' fishery.

40 **"There is nothing like them"**: Author interview, July, 2015.

41 **And just as humans bred canines**: E. H. Brown et al., "Historical Evidence for Discrete Stocks of Lake Trout in Lake Michigan," *Canadian Journal of Fisheries and Aquatic Sciences*, 1981; Charles Krueger and Peter Ihssen, "Review of Genetics of Lake Trout in Great Lakes: History, Molecular Genetics, Physiology, Strain Comparisons, and Restoration Management," *Journal of Great Lakes Research* 21, 1995.

42 **Strang, more latter day-pirate than saint**: *Milwaukee Sentinel*, September 1, 1940.

43 **And he became a self-styled naturalist**: James Strang, *Ninth Annual Report to the Board of Regents of the Smithsonian Institution*, 1855.

44 **Great Lakes fishing evolved from a local industry**: Margaret Beattie Bogue, *Fishing the Great Lakes: An Environmental History, 1793–1933* (Madison: The University of Wisconsin Press, 2000).

45 **"A few weeks ago Henry Smith"**: *Minneapolis Star Tribune*, as appeared in the Wakefield (Michigan) *News* on April 14, 1950.

46 **the lampreys had begun attacking**: *Corpus Christi Caller-Times*, December 28, 1950.

46 **The most ghastly thing**: Rob Gess, *Nature*, "A Lamprey from the Devonian of South Africa," October 26, 2006; author interviews with Gess and the *Nature* paper's co-author, Michael D. Coates of the University of Chicago.

51 **So a lamprey, nature-built to always swim upstream**: William Ashworth, *The Late, Great Lakes—An Environmental History* (New York: Alfred A. Knopf, 1986). I could find no research confirming this theory, which is the one favored by most historians on the subject, including John Burtniak, lamprey history expert and retired special collection librarian and university archivist at Brock University in St. Catherines, Ontario. Another possible explanation for the lamprey invasion above Niagara Falls is that they were accidentally planted by fishermen, or that there had long been a population in Lake Erie

following the opening of the first Welland Canal, but the population did not become large enough to be noticed until the 1920s.

52 **the first sea lamprey:** Vernon Applegate, "Sea Lamprey Investigations—An Inventory of Sea Lamprey Spawning Streams in Michigan," Michigan Department of Conservation, Fisheries Division Research Report 1154, 1948, 5.

52 **It would be another 15 years:** Associated Press, as reported in *Oshkosh Daily Northwestern*, May 30, 1936; "The Spread of the Sea Lamprey through the Great Lakes," Michigan Department of Conservation, Fisheries Division Research Report 381, 1936.

53 **"Unusually and exhaustively detailed":** Vernon Applegate, "Natural History of the Sea Lamprey, *Petromyzon marinus*, in Michigan," University of Michigan Institute for Fisheries Research, Report 1254, 1950.

54 **"living on cigarettes and aspirin":** Author interview, September, 2014.

55 **"Aquarium observations make it easily understandable":** Applegate, "Natural History of the Sea Lamprey," 223.

55 **"individuals of this life history state are seldom seen":** Ibid., 223.

55 **"When prodded":** Ibid., 95.

57 **"The male approaches the female":** Ibid., 148.

59 **"One of the most striking characteristics":** Ibid., 247.

60 **"the most vulnerable times in the lamprey's life":** Vernon Applegate and James Moffett, "The Sea Lamprey," *Scientific American* 192, no. 4, 1955, 37–41.

60 **experimental traps:** *Sheboygan Press Telegram*, December 1, 1950.

62 **"What I saw when I got here":** *Milwaukee Journal Sentinel*, December 4, 2010.

62 **The idea was to find a concoction:** "Toxicity of 4,346 Chemicals to Larval Lampreys and Fishes, 1957," U.S. Fish and Wildlife Service, Special Scientific Report—Fisheries, No. 207.

62 **Cliff Kortman's job:** *Milwaukee Journal Sentinel*, December 4, 2010.

63 **with "almost the secrecy of a nuclear project":** *Grand Rapids Herald*, May 25, 1958.

63 **"real purty sight":** Account published on November 13, 1957, in *The Brinewell*, a publication of the Dow Chemical Company.

64 **"the face of many a newly wolfless mountain":** Aldo Leopold, *A Sand County Almanac* (Oxford: Oxford University Press, 1949), 139–140.

65 **one stream in Maine alone:** Douglas Watts, *Alewife* (Augusta, Maine: Poquanticut Press, 2012), 65.

67 **there was little fanfare:** Robert Rush Miller, "Origin and Dispersal of the Alewife, *Alosa pseudoharengus*, and the Gizzard Shad, *Dorosoma cepedianum*, in the Great Lakes," *Transactions of the American Fisheries Society* 86, no. 1, 1957.

67 **"almost incredible" numbers:** Ibid.

67 **Their numbers at first were minuscule:** The Milwaukee River Technical Study Committee, *The Milwaukee River—An Inventory of its Problems, and Appraisal of its Potentials*, 1968.

68 **biologists estimated the river herring accounted for 17 percent of the fish mass:** Federal Water Pollution Control Administration, "The Alewife Explosion—the 1967 Die-off in Lake Michigan," July 25, 1967.

68 **They called them alewives:** The name alewife is also used on the East Coast, though alewives and blueback herring are commonly—and collectively—referred to as river herring.

69 **The pilot flying the Navy seaplane:** Ibid.

70 **if not the billions:** Ibid.

70 **There had been similar but smaller die-offs:** Ibid.

70 **they alone had disposed:** *Congressional Record*, Vol. 13, Part 14, June 29, 1967–July 18, 1967, 18927.

70 **"Chicago was running out of places":** Buffalo *Courier-Express*, July 9, 1967.

71 **"the fish are buried":** Federal Water Pollution Control Administration, "The Alewife Explosion—the 1967 Die-off in Lake Michigan," July 25, 1967.

71 **cleanup cost on Lake Michigan:** Oshkosh (Wisconsin) *Daily Northwestern*, September 22, 1967.

71 **costing the tourism industry:** Gerald R. Ford Presidential Library and Museum, February 29, 1968, news release while Ford was a member of the U.S. House of Representatives.

71 **six-foot-high mounds of carcasses:** *Chicago Tribune*, April 12, 1970.

71 **a single swimming mass of the fish:** The Milwaukee River Technical Study Committee, *The Milwaukee River—An Inventory of its Problems, and Appraisal of its Potentials*, 1968.

71 **as many as 6,000 fish:** Federal Water Pollution Control Administration, "The Alewife Explosion—the 1967 Die-off in Lake Michigan," July 25, 1967.

72 **"it was like hitting a snowbank":** *Milwaukee Journal Sentinel*, December 12, 2004.

72 **alewife breakfast sausages:** Associated Press, as appeared in the Lewiston (Maine) *Daily Sun*, October 6, 1975.

72 **"The stricken fish swam weakly":** "Population Characteristics and Physical Condition of Alewives, *Alosa Pseudoharengus*, in Massive Dieoff in Lake Michigan, 1967," Great Lakes Fishery Commission Technical Report No. 13.

73 **"The findings did not indicate any extreme or bizarre pollution conditions":** Federal Water Pollution Control Administration, "The Alewife Explosion—the 1967 Die-off in Lake Michigan," July 25, 1967.

74 "Lake Michigan was left with a fish population consisting largely of one species, the alewives": Ibid.

74 Applegate had a plan: Vernon Applegate and James Moffett, "The Sea Lamprey," *Scientific American* 192, no. 4, 1955, 37–41.

CHAPTER 3. THE WORLD'S GREAT FISHING HOLE

75 two businessmen from Waukegan: *Chicago Tribune*, March 9, 1969.

77 "the world's greatest fishing hole": Kristin M. Szylvian, "Transforming Lake Michigan into the 'World's Greatest Fishing Hole': The Environmental Politics of Michigan's Great Lakes Sport Fishing, 1965–1985," *Environmental History*, January 2004, 102–107.

77 "The salmon weren't put here to feed people": Author interview, January 2015.

78 the Tanners were not catch-and-release guys: All quotes in this chapter from Howard Tanner come from three interviews conducted with Tanner in fall 2014, some of which appeared in a *Milwaukee Journal Sentinel* series that ran December 7–9, 2014.

85 restock the Great Lakes with striped bass: Wayne Tody, *A History of Michigan's Fisheries* (Traverse City, Michigan: Copy Central, 2003).

88 "the introduction of foreign species": U.S. Department of Commerce—Bureau of Fisheries, *Fishing Industry of the Great Lakes*, 1925, 568.

89 "They didn't consult Canada": Author interview, January 2015.

89 "The coho is aimed": *Coho Salmon for the Great Lakes*, Michigan Department of Conservation, Fish Division, 1966.

92 an aerial tour of the upper Great Lakes: Wayne Tody, *A History of Michigan's Fisheries* (Traverse City, Michigan: Copy Central, 2003).

93 Towns in northwestern Michigan were swamped: The best account of this phenomenon relayed to me came in a 2004 interview with Jerry Dennis, author of the excellent *The Living Great Lakes: Searching for the Heart of the Inland Seas* (New York: Thomas Dunne Books, 2003). Dennis witnessed the phenomenon firsthand as a 13 year old. "There could have been hundreds of people killed," he told me.

93 "Motels filled for 50 miles around": "Coho Salmon Status Report," Fish Division of Michigan Department of Natural Resources, 1967–1968, 5.

93 "The one thing we did not fully anticipate was the fever": Associated Press, as reported in the Janesville (Wisconsin) *Daily Gazette*, October 17, 1968.

94 "All bodies eventually washed ashore": Ibid.

94 "This is Coho country": Oshkosh (Wisconsin) *Northwestern*, May 17, 1968.

94 **"Never before in fishery management"**: *Chicago Tribune*, March 8, 1968.

94 **jumped by $11.9 million**: "Coho Salmon Status Report," Fish Division of Michigan Department of Natural Resources, 1967–1968, 7.

95 **"There was so much pressure for us to follow"**: *Milwaukee Journal Sentinel*, December 12, 2004.

96 **"flash in the pan"**: Tody, *A History of Michigan's Fisheries*.

96 **"we may be raising alewives"**: Associated Press, as reported in Janesville (Wisconsin) *Daily Gazette*, October 17, 1968.

97 **the "scare" of the high DDT concentrations**: Associated Press, as reported in the *Evening News* (Sault Ste. Marie, Michigan), August 9, 1969.

98 **"Tanner got his dream"**: Author interview, January 2015.

98 **developing a super strain of salmon**: *Field and Stream*, December 1985, 79–84.

99 **a million of Tanner's "super salmon"**: Chicago *Tribune*, February 7, 1988.

100 **"Don't tell me the alewife don't eat the goddamn perch fry"**: *Milwaukee Journal Sentinel*, December 12, 2004.

101 **"It's just—it's just depressing"**: *Milwaukee Journal Sentinel*, December 7, 2014.

103 **"They had nothing to lose. The lakes were so destroyed"**: Ibid.

104 **"we can't control the Great Lakes"**: Ibid.

CHAPTER 4. NOXIOUS CARGO

108 **The first day of June 1988 was sunny**: All quotes and scene descriptions from Santavy come from interviews with the author in fall 2013 for an article in *Milwaukee Journal Sentinel*, July 27, 2014. Santavy also recounted her discovery of the first zebra mussel in *Quagga and Zebra Mussels, Biology, Impacts and Control* (Boca Raton, Florida: CRC Press, 2014), 3–4.

111 **Hungary succumbed to an infestation in 1794**: G. L. Mackie et al., "The Zebra Mussel *Dreissena polymorpha*: A Synthesis of European Experiences and a Preview for North America," report for the Ontario Ministry of Environment, 1989.

112 **"The *Dreissena* is perhaps better fitted"**: Early predictions of zebra invasions provided by James T. Carlton, emeritus professor of marine sciences, Williams College.

113 **A final warning came in 1981**: *Milwaukee Journal Sentinel*, December 26, 2004.

113 **"This guy hitchhiked inside a ballast tank"**: The Windsor (Ontario) *Star*, July 27, 1988, reproduced in the *Quagga and Zebra Mussels, Biology, Impacts and Control* (Boca Raton, Florida: CRC Press, 2014), 119.

114 **Editors at the Cleveland *Plain Dealer***: Cleveland *Plain Dealer*, August 29, 1868.

114 **You may think you know this story**: Jonathan Adler, "Fables of the Cuyahoga: Reconstructing a History of Environmental Protection," *Fordham Environmental Law Journal*, Faculty Publications, Paper 191, 2002. Adler makes the case that what is so significant about the 1969 fire is not that a river was set ablaze, but that this was the *last* time the Cuyahoga burned, and that the environmental condition of the river had actually hit its nadir decades earlier: "For northeast Ohio, and indeed for many industrial areas, burning rivers were nothing new, and the 1969 fire was less severe than prior Cuyahoga conflagrations. It was a little fire on a long-polluted river already embarked on the road to recovery."

David and Richard Stradling, "Perceptions of the Burning River: Deindustrialization and Cleveland's Cuyahoga River," *Environmental History* 13, no. 3, 2008. The authors contend the 1969 fire became such a touchstone because by the late 1960s fewer Clevelanders were employed by factories that polluted the river, and the further people moved from the polluted riverbank, the more absurd the degradation left behind became.

114 **Not much changed**: Cleveland *Plain Dealer*, November 2, 1952.

115 **"shriveling blast of blue flame"**: Cleveland *Plain Dealer*, May 2, 1912.

115 **"The Cuyahoga River was beautiful once"**: *Janesville* (Wisconsin) *Daily Gazette*, July 14, 1969.

116 **Federal environmental regulators**: UPI, as appeared in the *News-Journal* (Mansfield, Ohio), October 7, 1969.

116 **The U.S. Secretary of the Interior called for public hearings**: Ibid.

117 **Nobody was fined a dime**: I could find no historical evidence of any civil or criminal fines being levied for the fire, and neither has historian Jonathan Adler, a Case Western Reserve University School of Law professor who has researched and written extensively on the Cuyahoga River (email exchange with Adler, Aug. 7, 2015).

118 **"This type of discharge generally causes little pollution"**: *Federal Register*, May 22, 1973, 13,528, as cited in Senate Report 113-304, U.S. Government Publishing Office, December 10, 2014; Author interview with Nina Bell, executive director for the Northwest Environmental Advocates, which successfully sued to overturn the ballast water exemption; Daniel E. O'Toole, "Regulation of Navy Ship Discharges Under the Clean Water Act: Have Too Many Chefs Spoiled the Broth?" *William and Mary Environmental Law and Policy Review* 19, no. 1, 1994, 12; Jeff Alexander, *Pandora's Locks: The Opening of the Great Lakes—St. Lawrence Seaway* (East Lansing: Michigan State University Press, 2009).

119 **whether to call it a clam or a mussel**: A digital copy of the video tape of the 1989 conference was provided by conference host Ronald W. Griffiths.

120 "Nothing is going to be the same. Nothing": *Milwaukee Journal Sentinel*, July 27, 2014.

123 "People look at the lake and don't think of it as having a geography": Ibid.

126 Consider the alewife: "Status and Trends of Prey Fish Populations in Lake Michigan 2014," U.S. Geological Survey.

127 "I wouldn't say we've crashed": *Milwaukee Journal Sentinel*, October 10, 2015.

131 greatest catastrophe in the history of Great Lakes navigation: Much of the information on the sinking of the S.S. *Eastland* comes from the Eastland Disaster Historical Society: http://www.eastlanddisaster.org. Additional information comes from Michael McCarthy's excellent *Ashes under Water: The SS* Eastland *and the Shipwreck that Shook America* (Guilford, Connecticut: Lyons Press, 2014).

132 "Fred Swigert, a city fireman": *New York Times*, July 26, 1915. (There is no listing of a Swigert in the Eastland Disaster Historical Society database, though there is a listing of a Kathryn Swingert, no age available.)

134 "entire lack of understanding,": *New York Times*, August 12, 1915.

135 "destroying our Great Lakes": *Milwaukee Journal Sentinel*, October 31, 2005.

137 Jake Vander Zanden knows: *Milwaukee Journal Sentinel*, July 28, 2014.

139 Rudi Strickler, a Swiss-born zooplankton expert: Ibid.

142 "If you have only 500 fish": *Milwaukee Journal Sentinel*, July 29, 2014.

143 "This . . . is not rocket science": Ibid.

144 "I'd use the term 'ambitious,'": Ibid.

145 "Offload the cargo in Nova Scotia": *Milwaukee Journal Sentinel*, October 3, 2010.

145 "The solution could be simple": *Milwaukee Journal Sentinel*, July 29, 2014.

146 "It's a reasonable alternative, absolutely": Ibid.

CHAPTER 5. CONTINENTAL UNDIVIDE

154 The station wagon's tailgate was dropped: A picture of the arrival of the first Asian carp in the United States appeared in the federal government's *Draft Management and Control Plan for Asian Carps in the United States*, April 2006.

154 "When they did this, this was right. This was the thing to do": *Milwaukee Journal Sentinel*, August 19, 2012.

154 an Arkansas fish farmer: Much of the historical material in this chapter is derived from the personal papers of fish farmer and one-time Arkansas gubernatorial hopeful Jim Malone that are archived at the University of Central Arkansas.

155 "We had this little agreement": *Milwaukee Journal Sentinel*, October 15, 2006.

155 The fish farmer gave Henderson's: Malone papers, University of Central Arkansas.

155 they turned to S. Y. Lin, a professor at National Taiwan University: Ibid.

156 "we sent sample after sample [of fish] from the sewage ponds": *Milwaukee Journal Sentinel*, October 15, 2006.

158 "Redneck Fishing Tournament": *Milwaukee Journal Sentinel*, October 15, 2006.

158 "If these things get into the Great Lakes": Ibid.

159 Briney can catch 15,000 pounds of bigheads: *Milwaukee Journal Sentinel*, October 16, 2006.

161 "I'm old enough and big enough": *Milwaukee Journal Sentinel*, August 18, 2012.

163 There was little pomp in a ceremony: *New York Times*, January 17, 1900.

163 "Water in the Chicago River Now Resembles Liquid": *New York Times*, January 14, 1900.

164 "There is nothing which can be detected": U.S. Supreme Court, Missouri vs. Illinois, 200 U.S. 496, February 19, 1906.

166 "it is just a matter of time before we end up with a carp pond": *Milwaukee Journal Sentinel*, December 26, 2004.

166 "The night before, we'd get picnic baskets, beach balls": *Milwaukee Journal Sentinel*, August 19, 2012.

168 "I was one of those kids": *Milwaukee Journal Sentinel*, August 22, 2012.

172 "Shocked": Ibid.

176 "it's time to man the barricades": *Milwaukee Journal Sentinel*, December 4, 2009.

176 "I'm likening this action to chemotherapy": *Milwaukee Journal Sentinel*, November 13, 2009.

178 "we wouldn't have any positive samples above the barrier": *Milwaukee Journal Sentinel*, August 26, 2012.

179 sent him off with a dead Asian carp: *New York Times*' blog, The Caucus, October 1, 2010.

180 "They know what nets are . . . and they avoid them": *Milwaukee Journal Sentinel*, August 26, 2012.

183 "I'm pretty good at killing bills": *Milwaukee Journal Sentinel*, December 17, 2012.

184 "This is not about Asian carp": Ibid.

185 "Those aren't carp": *Milwaukee Journal Sentinel*, February 8, 2014.

185 "You can identify that they are fish": Ibid.

CHAPTER 6. CONQUERING A CONTINENT

187 **"The Great Lakes are just a beachhead"**: *Milwaukee Journal Sentinel*, November 2, 2009.

188 **densities exceeding 4,000 per square meter**: Amy J. Benson, "Chronological History of Zebra and Quagga Mussels (*Dreissenidae*) in North America, 1988–2010," a paper that appeared in *Quagga and Zebra Mussels: Biology, Impacts and Control* (Boca Raton, Florida: CRC Press, 2104), 11.

189 **a biologist for the Chicago sewerage district**: interview with Irwin Polls, former biologist for Metropolitan Water Reclamation District of Greater Chicago, January 2015.

189 **70 million per second**: Amy J. Benson, "Chronological History of Zebra and Quagga Mussels (*Dreissenidae*) in North America, 1988–2010," a paper that appeared in *Quagga and Zebra Mussels: Biology, Impacts and Control* (Boca Raton, Florida: CRC Press, 2104), 11.

190 **"I'm a nurse"**: *Milwaukee Journal Sentinel*, November 2, 2009.

191 **"If you want to know what's coming next"**: Ibid.

193 **"We didn't think much about it"**: *Milwaukee Journal Sentinel*, February 21, 2009.

194 **The pilot's job on July 21, 1948, was to dive that bomber from 30,000 feet**: *Smithsonian Magazine*, October 2005.

195 **"a little fingernail-sized mollusk"**: *Milwaukee Journal Sentinel*, February 21, 2009.

196 **"There is plenty of opportunity to control them in a pipe"**: Author interview, April, 2014.

198 **"That suggested a pretty significant risk"**: *Milwaukee Journal Sentinel*, July 29, 2014.

201 **"The reason I need to do this"**: The roadside exchange was captured on an audio recording made by the law enforcement officer. This and other materials related to the Regelman investigation were acquired with a request under Utah Government Records Access and Management Act filed with Kane County (Utah) in April 2014.

204 **"You can't remove every single mussel"**: *Milwaukee Journal Sentinel*, July 29, 2014.

204 **"The last thing they need . . . is another operation and maintenance cost or burden"**: *Milwaukee Journal Sentinel*, February 21, 2009.

207 **"I'll do the best I can"**: *Milwaukee Journal Sentinel*, July 29, 2014.

210 **"I think eDNA is emerging"**: Ibid.

CHAPTER 7. NORTH AMERICA'S "DEAD" SEA

212 **As there are lakes, and then there are the Great Lakes**: Three excellent accounts of the history of the Great Black Swamp are Howard Good, *Black Swamp Farm* (Columbus: Ohio State University Press, 1997); Jim Mollenkopf, *The Great Black Swamp: Historical Tales of 19th Century Northeast Ohio* (Toledo, Ohio: Lake of the Cat Publishing, 1999); and Martin R. Kaatz, *The Black Swamp: A Study in Historical Geography*, Annals of the Association of American Geographers, 1955.

213 **"There is a funeral every day"**: originally appeared in 1837 edition of the Maumee *City Express*, reproduced in *Commemorative Historical and Biographical Record of Wood County, Ohio: Its Past and Present* (Chicago: J.H. Beers and Co., 1897).

214 **"the wagon, getting off the beaten track"**: Account cited in Calvin Goodrich's *The Lie of the Land, Michigan Alumnus Quarterly Review, October 28 1944*. Published by the Alumni Association of the University of Michigan, 164.

215 **"as though it were the enemy"**: *The Story of the Great Black Swamp*, a television documentary written by Joseph A. Arpad and produced by WBGU-TV, 1982.

216 **"this shore which does not yet show any trace"**: George Pierson, *Tocqueville in America* (Baltimore: The Johns Hopkins University Press, 1938), 294.

216 **He is lucky he made the trip when he did**: *The Encyclopedia of Cleveland History*, a joint effort by Case Western Reserve University and the Western Reserve Historical Society, https://ech.case.edu

219 **Brand . . . with had little or no formal medical**: Several sources informed this section on phosphorus, none as much as John Emsley, *The Shocking History of Phosphorus, a Biography of the Devil's Element* (London: Pan Books, 2001).

222 **"The Lorax came out of my being angry"**: Lisa Lebduska, "Rethinking Human Need: Seuss's The Lorax," *Children's Literature Association Quarterly* 19, no. 4, 1994.

227 **"You'd get really sick"**: All quotes from Toledo and environs unless otherwise noted come from author interviews during a July 2014 trip to northwestern Ohio in the days leading up to the water shutdown for a story that subsequently appeared in the *Milwaukee Journal Sentinel*, September 13, 2014. One exception is quotes from Toledo Mayor Michael Collins during the height of the shutdown, which were culled from local television coverage.

229 **"Lake Erie today is not the same lake"**: *Milwaukee Journal Sentinel*, September 13, 2014.

241 **"My biggest fear"**: *Milwaukee Journal Sentinel*, September 24, 2014.

242 **"The canary in the coal mine"**: Ibid.

242 **"Without changes to the Act's approach"**: U.S. Government Accountability Office, "Clean Water Act: Changes Needed if Key EPA Program Is to Help Fulfill the Nation's Water Quality Goals," January 3, 2014.

243 **"I would love the pictures . . . to be forgotten"**: *Impact of Harmful Algal Blooms Requires Action*, Toledo Mayor Collins' testimony before the Senate Agriculture, Nutrition and Forestry Committee, December 3, 2014.

CHAPTER 8. PLUGGING THE DRAIN

247 **a "fuse" to the biggest carbon bomb**: Bill McKibben et al., *Grist*, June 23, 2011.

247 **"look silly"**: O.Canada.com, February 24, 2014.

249 **"Maybe I should pay a little more attention"**: *Milwaukee Journal Sentinel*, March 24, 2008.

250 **When Tennessee became the 16th state**: "History Corner: The Mystery of the Camak Stone," *Professional Surveyor Magazine*, March 2004.

251 **The survey team was given a simple directive**: Much of the history on Atlanta's founding comes from Skye Borden, *Thirsty City: Politics, Greed, and the Making of Atlanta's Water Crisis* (Albany: SUNY Press, 2014).

253 **"beyond our borders"**: Associated Press, October 20, 2007, as reported in the *Washington Post*.

253 **"It's like having a good wife"**: *Milwaukee Journal Sentinel*, March 23, 2008.

254 **"I get lots of calls that start with 'bitch'"**: Ibid.

255 **"What I thought was a joke"**: The (Chattanooga, Tennessee) *Times Free Press*, as reported in *Smithsonian.com*, September 26, 2012.

255 **"We are not going to move the state line"**: The (Chattanooga, Tennessee) *Times Free Press*, February 18, 2014.

255 **"the kind of conflict that 'might lead to war'"**: C. Crews Townsend, "Crossing the Line—Does Georgia Plan to Redraw the Tennessee–Georgia Border Pass Legal Muster?" *Tennessee Bar Journal*, June 20, 2008.

256 **an 18-year-old boy with a beak**: *The History of Waukesha County, Wisconsin* (Chicago: Western Historical Company, 1880), 493–496.

259 **Dunbar, as he told the story near his life's end**: *The History of Waukesha County, Wisconsin*, 330–333; "Spring City and the Water War of 1892," *The Wisconsin Magazine of History* 89, no. 1, Autumn 2005.

261 **"Now began the 30 years of the Saratoga of the west"**: *Milwaukee Journal*, June 6, 1953; for an exceptionally detailed account of the history of Waukesha's spring era, see John Schoenknecht, *The Great Waukesha Springs Era: 1868–1918* (J. M. Schoenknecht, 2003).

261 **"sufficient to more than slake the thirst"**: *The History of Waukesha County, Wisconsin*, 328.

262 **Chicago health records from 1891 reveal typhoid**: Dr. Bronwyn Rae, "Water, Typhoid Rates, and the Columbian Exposition in Chicago," *Northwestern Public Health Review* 3, no. 1, 2014; Michael P. McCarthy, "Should We Drink the Water? Typhoid Fever Worries at the Columbia Exposition," *Illinois Historical Journal*, Spring 1993.

262 **"The new corporation to fill its contract"**: *Chicago Daily Tribune*, May 9, 1892.

263 **McElroy, carrying a pistol and $8,000 in cash**: *Waukesha Daily Freeman*, May 12, 1892.

263 **"Every one thought that something terrible"**: *Chicago Daily Tribune*, May 9, 1892.

263 **McElroy eventually did get**: *Milwaukee Journal Sentinel*, September 4, 2006.

263 **no reported typhoid deaths**: Dr. Bronwyn Rae, "Water, Typhoid Rates, and the Columbian Exposition in Chicago," *Northwestern Public Health Review* 3, no. 1, 2014.

265 **"You go and look at Lake Superior, and you say, 'Look at all that water'"**: "Inside Rollingstones, Inc.," *Fortune Magazine*, September 30, 2002.

266 **"Trying to describe what the Aral Sea is"**: *Milwaukee Journal Sentinel*, November 7, 2006.

268 **"The south has plenty of water and the north lacks it"**: Obja Borah Hazarika, "Riparian Relations between India and China: Exploring Interactions on Transboundary Rivers," *International Journal of China Studies* 6, no. 1, April 2015, 63–83.

269 **in 2002 the Chinese government broke ground**: *New York Times*, June 1, 2011.

269 **"Western states and Eastern states have not been talking"**: *Milwaukee Journal Sentinel*, October 6, 2007.

270 **"the state has only about one year of water supply left"**: *Los Angeles Times*, March 12, 2015.

271 **"You can't have major cities and civilizations where there is not adequate water"**: Author interview, January 2016.

271 **environmental commissioner said it was "inescapable"**: *Buffalo News*, September 12, 1985, as cited in Peter Annin, *Great Lakes Water Wars* (Washington, D.C.: Island Press, 2006), 78; *New York Times*, August 28, 1985.

272 **triple flows through the Chicago canal**: *New York Times*, July 1, 1988.

273 **"I was doing it for a purpose"**: Interview with author, October 13, 2015. For a more detailed account of the plan to ship away the Great Lakes, see Annin's *Great Lakes Water Wars*.

273 **"This is Pandora's Box"**: Associated Press, as appeared in the Ludington (Michigan) *Daily News*, May 1, 1998.

274 **"to be ideologically so out of touch?"**: Associated Press, as appeared in the *Toledo Blade*, September 18, 2005.

276 **"We are not opposed to a Waukesha diversion"**: Author interview, January 2016.

CHAPTER 9. A SHAKY BALANCING ACT

All quotations in this chapter, unless otherwise noted, come from a two-day series on climate change and lake levels that ran in the *Milwaukee Journal Sentinel* on July 27 and July 30, 2013.

286 **It was dead calm**: *Annual Reports of the War Department for the Fiscal Year Ended June 30, 1902*, Government Printing Office, 2244.

286 **As the *Fontana* approached this roiling hole**: *The Federal Reporter*, Vol. 119: Cases Argued and Determined in the Circuit Court of Appeals and Circuit and District Courts of the United States, February–March, 1903, 856–863.

287 **the *Fontana* collided**: *Detroit Free Press*, Sept. 22, 1900; *The Federal Reporter*, Vol. 117: Cases Argued and Determined in the Circuit Courts of Appeals and Circuit and District Courts of the United States, October–November 1902, 894; *Chicago Tribune*, September 22, 1900.

288 **One dredging project**: *International Upper Great Lakes Study, Impacts on Upper Great Lakes Water Levels: St. Clair River, Final Report to the International Joint Commission*, December 2009.

289 **The team built a miniaturized southern Lake Huron and St. Clair River**: *Effects of Submerged Sills in the St. Clair River, Hydraulic Model Investigation*, U.S. Army Corps of Engineer Waterways Experimental Station Technical Report H-72-4.

291 **"It should be remembered"**: Ibid.

293 **"We've got something alarming going on here"**: *Milwaukee Journal Sentinel*, January 25, 2005.

299 **"For the moment"**: Author interview, April 2015.

CHAPTER 10. A GREAT LAKE REVIVAL

300 **Ken Koyen is the last full-time commercial fisherman**: I have interviewed Koyen more than a dozen times over the past decade, the first time for a *Milwaukee Journal Sentinel* story that appeared on April 3, 2003.

301 **"Honestly, I thought whitefish were done"**: *Milwaukee Journal Sentinel*, August 16, 2011.

302 **"Look at that! Look at that!"**: Ibid.

302 **"They're fat"**: Ibid.

303 "'No way Charlie. Stop the bullshitting'": Ibid.

304 "the fish can still thrive": Ibid.

304 "more like a snake than a fish": *Milwaukee Journal Sentinel*, June 30, 2008.

304 "As you're driving through town": Ibid.

305 "It all started as soon as the alewives were gone": *Milwaukee Journal Sentinel*, December 9, 2014.

306 "Those species that can make the switch to gobies are OK": Ibid.

306 "You wanted the alewife to be alive and healthy": Ibid.

306 "If you really want a native fish recovery": Ibid.

307 "We never had a boat": Ibid.

308 "The salmon are from the Pacific": Author interview aboard the *R/V Arcticus*, September 2015.

310 "a laundry list of why this can't be done": *Milwaukee Journal Sentinel*, February 8, 2014.

311 "There is a deficiency of governance here": *Milwaukee Journal Sentinel*, October 25, 2009.

313 Ron Thresher, an American who pioneered: Most of the material from Tasmania comes from a November 2010 trip to Tasmania to research the DNA technology for a story that appeared in the *Milwaukee Journal Sentinel* on December 4, 2010.

314 "a single carrier of one of these could doom a population": Author interview, April 2015.

315 "To have the potential to rectify our past environmental mistakes": Author interview, April 2015.

315 "The basic technology, once it's up and running": *Milwaukee Journal Sentinel*, December 4, 2010.

316 "We are not even close to developing a governance system": Author interview, January 2015.

316 "A thing is right": *A Sand County Almanac*, Aldo Leopold (Oxford University Press, 1949), 262.

317 "I doubt if the charter boat captains": *Milwaukee Journal Sentinel*, December 9, 2014.

317 "We want to do the same thing that Tanner did": Ibid.

318 "There is a body of evidence": Ibid.

318 "If you could smell the paper mill": Ibid.

320 "The idea is that you keep enough alewives": Ibid.

320 "The problem . . . is you guys are all old": Ibid.

SELECTED BIBLIOGRAPHY

Alexander, Jeff. *Pandora's Locks: The Opening of the Great Lakes—St. Lawrence Seaway*. East Lansing: Michigan State University Press, 2009.

Annin, Peter. *The Great Lakes Water Wars*. Washington D.C.: Island Press, 2006.

Ashworth, William. *The Late, Great Lakes: An Environmental History*. New York: Knopf, 1986.

Bernstein, Peter. *Wedding of the Waters: The Erie Canal and the Making of a Great Nation*. New York: W. W. Norton, 2005.

Bogue, Margaret Beattie. *Fishing the Great Lakes: An Environmental History, 1783–1933*. Madison: University of Wisconsin Press, 2000.

Bordon, Skye. *Thirsty City: Politics, Greed, and the Making of Atlanta's Water Crisis*. Albany: State University Press of New York, 2014.

Chiarappa, Michael, and Kristin Szylvian. *Fish for All: An Oral History of Multiple Claims and Divided Sentiment on Lake Michigan*. East Lansing: Michigan State University Press, 2003.

Crawford, Stephen. *Salmonine Introductions to the Laurentian Great Lakes: A Historical Review and Evaluation of Ecological Effects*. National Research Council of Canada, 2001.

Dempsey, Dave. *Ruin & Recovery: Michigan's Rise as a Conservation Leader*. Ann Arbor: The University of Michigan Press, 2001.

Dempsey, Dave. *On the Brink: The Great Lakes in the 21st Century*. East Lansing: Michigan State University Press, 2004.

Dennis, Jerry. *The Living Great Lakes: Searching for the Heart of the Inland Seas*. New York: Thomas Dunn Books, 2003.

Emsley, John. *The Shocking History of Phosphorus*. London: Macmillan, 2000.

Fischer, David Hackett. *Champlain's Dream*. New York: Simon and Schuster, 2008.

Good, Howard. *Black Swamp Farm*. Columbus: Ohio State University Press, 1997.

Grover, Velma, and Gail Krantzberg (eds.). *Great Lakes: Lessons in Participatory Governance*. Boca Raton, Florida: CRC Press, 2012.

Hartig, John. *Burning Rivers. Revival of Four Urban-Industrial Rivers*. Essex, United Kingdom: Multi-Science Publishing, 2010.

Hansen, Michael, and Mark Holey. "Ecological Factors Affecting the Sustainability of Chinook and Coho Salmon Populations in the Great Lakes." In *Sustaining North American Salmon: Perspectives across Regions and Disciplines*, edited by Michael Jones, Kristine Lynch, and William Taylor, 155–179. Bethesda, Maryland: American Fisheries Society, 2002.

Halverson, Anders. *An Entirely Synthetic Fish: How Rainbow Trout Beguiled America and Overran the World*. New Haven, Connecticut: Yale University Press, 2010.

Hill, Libby. *The Chicago River: A Natural and Unnatural History*. Chicago: Lake Claremont Press, 2000.

Hills, T. L. *The St. Lawrence Seaway*. New York: Frederick A. Praeger, 1959.

Hubbs, Carl, and Karl Lagler (revised by Gerald Smith). *Fishes of the Great Lakes*, revised ed. Ann Arbor: University of Michigan Press, 2007.

Keller, Reuben, Marc Cadotte, and Glenn Sandiford (eds.). *Invasive Species in a Globalized World: Ecological, Social & Legal Perspectives on Policy*. Chicago: University of Chicago Press, 2015.

Kuchenberg, Tom. *Reflections in a Tarnished Mirror: The Use and Abuse of the Great Lakes*. Sturgeon Bay, Wisconsin: Golden Glow Publishing, 1978.

Lesstrang, Jacques. *Seaway: The Untold Story of North America's Fourth Seacoast*. Seattle: Superior Publishing Company, 1976.

Macfarlane, Daniel. *Negotiating a River: Canada, the US, and the Creation of the St. Lawrence Seaway*. Vancouver: University of British Columbia Press, 2014.

Mabee, Carlton. *The Seaway Story*. New York: Macmillan, 1961.

McCarthy, Michael. *Ashes under Water: The SS Eastland and the Shipwreck that Shook America*. Guilford, Connecticut: Lyons Press, 2014.

Mollenkopf, Jim. *The Great Black Swamp: Historical Tales of 19th Century Northeast Ohio*. Toledo, Ohio: Lake of the Cat Publishing, 1999.

Nalepa, Thomas, and Don Schloesser (eds.), *Quagga and Zebra Mussels: Biology, Impacts, and Control*. Boca Raton, Florida: CRC Press, 2014.

Parham, Claire Puccia. *The St. Lawrence Seaway and Power Project: An Oral History of the Greatest Construction Show on Earth*. Syracuse: Syracuse University Press, 2009.

Pierson, George. *Tocqueville in America*. New York: Oxford University Press, 1938.

Ryan, William, and Walter Pitman. *Noah's Flood: The New Scientific Discoveries about the Event that Changed History*. New York: Simon & Shuster, 1998.

Scavia, Donald, et al. "Assessing and Addressing the Re-eutrophication of Lake Erie: Central Basin Hypoxia," *Journal of Great Lakes Research* 40, no. 2, 2014.

Shaw, Ronald. *Canals for a Nation: The Canal Era in the United States 1790–1860*. Lexington: The University Press of Kentucky, 1990.

Stagg, Ronald. *The Golden Dream: A History of the St. Lawrence Seaway*. Toronto: Dundurn Press, 2010.

Watts, Douglas. *Alewife: A Documentary History of the Alewife in Maine and Massachusetts*. Augusta, Maine: Poquanticut Press, 2012.

Willoughby, William. *The St. Lawrence Waterway: A Study in Politics and Diplomacy*. Madison: University of Wisconsin Press, 1961.

ILLUSTRATION CREDITS

40 © Emily S. Damstra

50 © Emily S. Damstra

56 © George Skadding/The LIFE Picture Collection/Getty Images

59 Courtesy of the NOAA Great Lakes Environmental Research Laboratory

68 © Emily S. Damstra

121 Courtesy of the Milwaukee Journal Sentinel

133 Photo courtesy of the Eastland Disaster Historical Society

157 Courtesy of the Milwaukee Journal Sentinel

196 Courtesy of the Milwaukee Journal Sentinel

200 Courtesy of the Milwaukee Journal Sentinel

303 Courtesy of the Milwaukee Journal Sentinel

322 Courtesy of the Milwaukee Journal Sentinel

INDEX

Note: Page numbers in *italics* refer to illustrations.
Page numbers after 325 refer to notes.

agricultural waste, xviii, 221, 222, 227–28, 231–32, 234–37, 243, 311
agriculture:
  crop buffers in, 233
  farmers working in, 232–36
  nonpoint pollution by, 231–32, 311
  no-till growing, 227–28
  nutrient management in, 236
  Ogallala Aquifer drained for, 267–68
  and rainfall, 228, 229, 237
  soil consultants in, 235–36
  variables in, 232
Alabama, water supply, 254
alewives, 38, 65–68, *68*
  control of, 104
  die-offs of, 69–74, 75, 83–84, 92, 96, 97–98
  on East Coast, 65, 66, 67–68
  as fertilizer, 65, 72, 95–96, 100
  in food chain, 65–66, 67–68, 76, 93, 103, 105, 157, 304–7, 319
  as invasive species, 67–68, 83, 135, 305, 306, 318–19
  and lake trout, 305–6
  migration of, 66–67, 76
  native species as prey of, 68, 99, 100, 305, 306
  population crashes of, 69–70, 105, 107, 126–27, 306, 308, 319
  protection of, 99–100, 318

  as salmon prey, 93, 95, 97, 99, 103, 106, 127, 304–7, 308, 318
  surveys of, 125, *126–27*
  as whitefish prey, 302
Alexander, Jeff, 327, 335
algae:
  blue-green, 224–25, 226
  and dead zones, 218–19, 224
  and eutrophication, 221–22
  in food chain, 160, 218
  functions of, 224
  HABs (harmful algae blooms), 239
  invasive species, 109–10
  invisible toxins from, 238
  Lyngbya, 226, 230
  shrinkage of, 223, 242–43
  Thalassiosira weissflogii, 109–10
  toxic outbreaks, xvii, xviii, 216, 224–31, 234–35, 236, 238, 239, 296
  uncontrollable, 110
Alicia Rae, 303
alum, 239
Ambs, Todd, 145
Anderson, Mark, 198–99, 205, 211
Annin, Peter, 183–84, 266
Apalachicola oyster beds, 254
Applegate, Vernon:
  and lake trout, 74, 84–85
  and sea lampreys, 53–61, 223
  testing poisons, 62–63, 83

aquatic insects, 39, 41, 137
Aral Sea, 265–66, 274
Arcticus, 125–27
Arizona:
  invasive species in, 195, 199, 207
  mussel possession illegal in, 198
Arkansas Game and Fish Commission, 155
Army Corps of Engineers, U.S.:
  Chicago canal plan (2014) of, 183, 309–10
  and electrical fish barrier, 165, 166–67,
    171, 174, 180, 185
  and erosion problems, 293–96
  and lawsuits, 178, 181–83
  and Ogallala Aquifer, 268
  and St. Clair River project, 289–91,
    293–96
  shipping industry served by, 290, 309–10
  and St. Lawrence Seaway, 30, 32, 109
  and water levels, 293–99
Asian carp:
  bighead carp, 154, 155–61, 157, 176–77,
    180–81, 182, 185–86, 310
  black carp, 154, 155, 161, 204–5, 310
  DNA testing for, 170–75, 177, 184
  ecological damage caused by, 310, 315
  and electrical barrier, 165–68, 171–75,
    176–81, 184, 185
  in government experiments, xvii, 154–61
  grass carp, 154
  in private fish pond, 185–86
  release of, 156, 160–61, 186
  and research tools, 182–83
  silver carp, 154, 155–61, 170, 176, 310
  sinking when poisoned, 182, 186
  spread of, 160–61, 167, 169, 174, 176, 178,
    179–81, 185, 309, 313
  in YouTube videos, 157–58, 178
Atlanta, Georgia, public water supply,
    250–56, 271
Atlantic Ocean, Great Lakes linked to, 38,
    108, 151, 265
atom bomb testing, 285
Auburn University, 155, 312
Austin, Jay, 282–84
Australia, and daughterless gene, 312–15

ballast water, 111–13
  and Clean Water Act, xv–xvi, 117–18, 141
  diseases traceable to, 144
  disinfection systems for use in, 136,
    138–39, 140–44, 308
  dumping of, xiv-xvi, 141
  and *Eastland* disaster, 131–34
  fines applied for infractions, 210–11
  flushing tanks in mid-ocean, 136, 143,
    208–9, 309

inadequate discharge rules for, 141
living organisms in, xv–xvi, xvii, 35, 109,
    131, 134–36, 144–45, 147, 169, 188,
    308–9, 311
  as national problem, xvii, 308
  residual, 134, 141
  salinity tests of, 209–10
  treatment standards for, 308, 311
bass:
  smallmouth, 39, 306, 317
  as sport fish, 81
  striped, 85, 207, 211
bateaux, cargo carried via, 12
Biel, Mark, 183
Biggert, Henry, 237–38, 239
birds:
  botulism outbreaks in, 130–31, 135
  food sources of, 66
bison, slaughter of, xviii–xix
Blackfoot River, 81
Black Sea:
  invasive species from, 127, 135, 187
  and the Mediterranean, 6–7
Black Sea Deluge, 6–7
Black Sea sturgeon, 6
bloody red shrimp, xvii, 135
Bootsma, Harvey, 123, 128–30
Bosporus Strait, 7
Bosporus Valley, 6
Boston Society of Natural History, 113
botulism, 130–31, 135
Brand, Henning, 219
Brazil, microsystis in, 224
Bridgeman, Tom, 225–30, 231, 232
Briney, Orion, 159–60
Brockel, Harry C., 23
Brown, Jerry, 270
Buckeye Traction Ditcher, 215
Buerkle, Robert, 267–68
Buffalo, New York, 17, 19
bullfrogs, American, 170
Burcham, Margaret, 185
Burggraaf, Kent, 202–3
Burtniak, John, 330
Bush, George W., 275

California:
  agricultural produce in, 270
  drought in, 247, 250, 270
  invasive species in, 195, 199, 207
  mussel possession illegal in, 198
Camak, James, 251, 255
Canada:
  and ballast water regulation, 145
  and domestic shipping, 146
  English capture of, 11

and GRAND Canal, 267
and hydroelectric power, 280
and international shipping, 145
and Keystone XL pipeline, 247
ore deposits of, 21
and plan to divert water to Asia, 274, 275
and St. Lambert Lock, 144–45
and St. Lawrence Seaway, 22, 25, 310
water diverted into Great Lakes from,
287
Welland Canal built by, 19–20
canals:
and commerce, 12–13, 19–20
ecological troubles caused by, 65
global trade in, 24, 31, 198, 309
invasive species moving in via, 45, 48, 49,
50, 111, 164–65, 189–90
navigation locks of, 11–12, 16, 19–20,
24–25, 173, 174–75, 178, 279
and navigation routes, 152–53
winter shutdowns of, 20
see also specific canals
Carnegie Mellon University, Remaking
Cities Institute, 270–71
carp:
Asian, *see* Asian carp
and daughterless gene, 312–15
and food chain, xvii, 154, 156–57, 310
hatchery program for, 88, 313–14
as invasive species, 135, 154–61, 313
Carroll Water and Sewer District, 237–38,
240
Carson, Rachel, 153
Carter, Donald K., 270–71
Cartier, Jacques, 8–9, 10
Caspian Sea, invasive species from, 127, 135,
187, 191
Castle, Lewis G., 35
catfish, 165, 310
Chadderton, Lindsay, 178, 184
Chapman, Duane, 180, 186
Chemical Industry Council of Illinois, 183
Chicago:
canal, *see* Chicago Sanitary and Ship
Canal
and commerce, 19, 24, 32
development of, 152–53
drinking water of, 163, 261–62, 263, 271
in nineteenth century, 257–58
Sears (Willis) Tower in, 264
sewage system of, 161–64, 183, 189, 261,
271, 296
typhoid in, 262, 263
World's Fair (1893) in, 261–62, 263, 275
Chicago River, 152, 162–63, 175, 179–80,
184, 261

Chicago Sanitary and Ship Canal, 161–66,
162
Army Corps of Engineers plan (2014),
183, 309–10
cargo movement on, 174
connection between Great Lakes and
Mississippi basin, xvii, 122, 151–53,
162–64, 184, 188–89, 271, 309
electrical fish barrier in, xvii, 165–68,
171–75, 176–81, 311
expansion of, 162
fish poisoned in, 175–78, 179–80, 182,
185–86
as Great Lakes "back door," xvii, 152, 153,
187, 309
invasive species spreading on, xvii–xviii,
122, 165, 188–89, 205, 309
lawsuits related to, 163–64, 181–83, 271,
280
natural barrier destroyed by, 181–84
original purpose of, 161–62
water diversion of, 163, 261, 280, 287
and water shortages, 272
China:
carp shipped to, 159
water diversion project in, 268–69
Chinese mitten crabs, 144
chlorine, 239
cholera, 144
Chrysler Corporation, 23
chubs, 100, 125, 301
ciscoes, 39, 41, 44, 45
Cladophora, 128, 130
clams, invasive, 144
Clean Water Act (1972), xiv–xvi, 114
and Cuyahoha River fires, 117, 147
and farm runoff, 231–32, 311
lawsuits related to, 138, 141
loopholes in, xvi, 117–18, 131, 134, 143,
231
and nonpoint pollution, 231–32, 311
passage of, 117, 308
positive effects of, 117, 242
and treatment technologies, 141, 143
Cleveland, factories in, 116–17, 216–17
climate change:
adaptive management of, 297–99
and air temperatures, 284–85
effect on water levels, xviii, 277, 279–80,
285–86
and erosion, 293
hydrologic regime, 285
and polar vortex, 296
and radio-carbon dating, 278
and water precipitation-evaporation cycle,
279, 294–95

climate change (*continued*)
  and water temperatures, 283–84, 285
  and weather cycles, 296–99
Clinton, Bill, 168
Clinton, DeWitt, 15–17
Coast Guard, U.S.:
  and ballast management, 131, 134
  and electrical fish barrier, 166, 167
  ice breakers of, 32
  primary mission of, 166
coho (silver) salmon, 76–78, 86–100
  alewives as prey of, 93, 97
  hatchery-bred, 76, 91
  jack males, 92
  in Lake Michigan, 76, 91–94
  life cycle of, 91, 99
  migration of, 92–94
  for sports fishing, 77–78, 89, 93–94
  Tanner's introduction into Great Lakes of,
    86–94, 96–100, 103–4
Collins, Michael, 240–42, 243–44
Collins Park Water Treatment Plant, Toledo,
  238–40
Colorado, invasive species in, 195
Colorado River, 192–97
  canal system of, 197
  dams on, 193, 270
  and Grand Canyon, 286–87
  mussels in, 193–95, 197
  plans to suck it dry, 192–93
  water shortage on, 270
Columbia River:
  efforts to keep mussels out of, 208
  hydroelectric dam system of, xviii, 197
Columbus, Christopher, xii
containerships, 26–27, 32–33
continental divide, 151–52
corn, used in ethanol production, 243
Creaser, Charles, 46
Cronkite, Walter, 3–4
crustaceans, 39, 41, 110
Cudahy, Brian J., 26
cultural eutrophication, 218
Cuyahoga River:
  and Clean Water Act, 117, 147
  on fire, xiv, xvi, 114–16, 147, 217, 232, 242
  fish in, 117
  media stories about, 114, 115
  as warning, 243
Cuyahoga Steam Furnace Co., 216

Dailey, Candy, 190, 191
Daley, Richard, 145
dams, manmade water bodies behind, 81
daughterless gene, 312–15
Davis, Cameron, 176

Davis Dam, 195
DDT, 97
dead zones, 218–19, 224
Deaton, Janice, 304
Delaware River, Washington's crossing, 12
Dennis, Jerry, 333
Des Plaines River, 152, 163
Detroit:
  and commerce, 19
  wastewater from, 234
Detroit River, 234–35
Dettmers, John, 318
Diamond Valley Lake Reservoir, 196–97
dinoflagellates, 136
DNA:
  and daughterless gene, 312–15
  as eradication tool, 316
  as evidence, 169–70, 177, 182
  genetic markers (primers), of, 170, 209
  testing for, 169–75, 176, 177–78, 179, 184,
    199, 205, 209–10
Doer, Gary, 247
Door Peninsula, Wisconsin, 184, 303
Douglas, Charles, 43
drain tiles, 215, 236–37
Dreyfuss, Richard, 225
Duluth, and commerce, 32, 33
Dunbar, Richard, 259–60
Dunham, Rex, 314–15
Durham boats, cargo carried by, 12
Dust Bowl, 280

Eastland, S.S., 131–34, *133*
E. coli, 312
ecosystems:
  affected by draining the lakes, 266–67
  artificial management of, 81, 99–100
  role of predators in, 64–65, 67, 84
  ruined by invasive species, 88, 123, 135,
    146, 153, 156–57, 229, 310, 315
Eder, Tim, 295, 297
Egan, John, 321, 322
Eisenhower, Dwight D., 21–22, 27–28
electrical fish barriers, 60–61, 165–68,
  171–75, 176–81, 184–85, 311
Elizabeth II, queen of England, 27–28
Emanuel, Rahm, 179, 242
emerald shiner, 38
Endangered Species Act, 66
Environmental Protection Agency (EPA):
  ballast water exemption of, xvi, 131, 311
  ballast water regulation by, 138–39, 140–
    44, 145, 308, 311
  citizen lawsuits against, 138, 141
  and Clean Water Act, xv, 117–18, 131, 138,
    141, 143, 308

and invasive species, 138, 311
and restoration programs, 311
and sewage treatment experiments, 155–56
Erie Canal:
  and commerce, 18–20
  construction of, 38
  expansion of, 20
  and Hudson River, 15–17
  impact of, 18–19
  and invasive species, 48, 49
  and Lake Erie, 15–18
  made obsolete, 20
  opening of, 48
  and water supplies, 271–72
  and western expansion, 15, 17
  winter shutdowns of, 20
Eurasian ruffe, 113
Europe:
  commerce from Great Lakes to, 20
  invasive species in, 111, 112
European flatworm, 110
eutrophication, 218, 221–22
Evinrude, 94
Exxon Valdez, xvi–xvii

Fahnenstiel, Gary, 146, 229
Famiglietti, Jay, 270
Farden, Art, 304–5
Febbraro, John, 272–74, 294
Federal Water Pollution Control
    Administration, 69, 71
fertilizer:
  alewives as, 65, 72, 95–96, 100
  costs of, 235
  in no-till growing, 227–28
  overapplication of, xviii, 236
  reductions of, 232, 235
  runoff of, xviii, 221, 222, 227–28, 231–32,
    234–37, 243, 296, 311
  seasonal application of, 228
  and soil testing, 235
Fielder, Dave, 103–4, 306
Fiesta Queen, 200–202
fish:
  artificial food for, 87
  artificially stocked, 76, 77, 81, 90–96,
    98–100, 102–4, 135, 154–55, 316
  baitfish, 99, 191
  decimated populations of, 45, 69–74,
    124–27, 308, 318
  DNA testing for, 170–75
  ear bones of, 181
  in food chain, 38–39, 41, 49, 113, 125, 318
  habitat destruction of, 86
  hatchery-bred, 76, 77, 81, 87, 95, 104–7,
    197, 313–14

and industrial pollution, 96–97
invasive species, 191; see also alewives;
    carp
native species, 77, 83, 88, 96, 99–100,
    301, 302–3, 305–8
origins of, 36–37
surveys of, 125–27, 182–83, 308
unsustainable numbers of, 102
upstream migration of, 38
and water levels, 295
see also specific fish
Fish Farming Experimental Laboratory,
    154–55
fish farms, and invasive species, 154, 160,
    310, 313
fishhook water flea, xvii, 135
fishing industry:
  commercial harvests, 82, 84, 300–302
  evolution of, 44
  for fertilizer, 95–96
  and fishery management, 82–83, 84, 100,
    104, 107
  harvest yields in, 44
  lamprey's impact on, 53, 64, 65
  and native species, 77
  overfishing by, 67, 86
  and sports fishing, see sports fishing
  fish-killing virus, 165
Florida:
  pythons in, 315
  water supply, 254
Fontana, 286–87, 298
Food and Agriculture Organization (FAO),
    155
Food and Drug Administration (FDA), 156
Fortner, Rosanne, 223
"Fourth Seacoast," 8, 11, 18, 30
Francis I, king of France, 8–9
Franklin, Benjamin, 253
Franklin, Shirley, 252–53
Freeze, Mike, 155–56, 160
freighters:
  ballast water of, see ballast water
  and canal locks, 19–20, 279
  and Clean Water Act, xvi, 118, 131
  containerships, 26–27, 32–33
  contaminated water discharged from,
    xiv–xvi, 35, 111, 308
  domestic, 146–47
  elevator system for, 28
  "lakers," 188
  and navigational dredging, 288–89
  offloading cargo at point of entry, 145
  and St. Clair River, 288–89, 292
  "salties," 145–46
  sludge in ballast of, 134

freighters (*continued*)
  and St. Lawrence Seaway, xv, 21, 30, 109,
    113, 145–47, 187–88, 191, 208–11
  total tonnage carried by, 146
  vast range of, 144
  see also shipping industry
freshwater:
  balanced "immune system" of, 10
  evolution of, 10–11
  in Great Lakes, xi, xiii, 82, 248, 249, 265
Frost, Leslie, 21

Gary, Indiana, 217
Geisel, Theodor S. (Dr. Seuss), 217, 222–24
Georgia:
  droughts in, 254, 275
  railroads in, 251, 252, 256
  state line (35th parallel) of, 250–56
Georgian Bay study, 291–94
Gess, Rob, 47–49
Gingrich, Newt, 274
glaciers:
  and erosion, 293
  and fish, 36
  "snout" of, 37
Glen Canyon Dam, 206–7
gobies, 127, 128, 130, 135, 302, 306, 320
Gondwana, 47, 48
Government Accountability Office, 242, 311
Grace Line, 29
GRAND (Great Recycling and Northern
    Development) Canal, 267
Grant, Ulysses S., 261
Gray, Lorri, 204–5
Great Black Swamp, Ohio, 212–16, 214, 257
  author's quest for information on, 225–32
  farming practices in, 228, 233
  natural water purification in, 231
Great Lakes:
  artificially stocked fish in, 90–96, 98, 316
  balance sought in, xviii, 106, 107
  botulism outbreaks in, 130–31, 135
  control points for outflow of, 279–80
  draining water from, xviii, 163–64, 248–
    76
  "ecologically naïve" period of, 11, 49, 53
  economic costs of invasive species in, 197
  ecosystem destroyed, 53, 68, 103, 146
  erosion effects on, 292–95, 298
  European explorers of, 8–10
  fish of, 36, 44; *see also* fish; fishing indus-
    try; *specific fish*
  freighters on, *see* freighters
  freshwater in, xi, xiii, 82, 248, 249, 265
  and Gulf of Mexico, 152, 153, 162
  hydrologic cycles, 297–99

ice age evolution of, 37–38
  ice cover on, 281–82, 283–84, 296
  impact of Erie Canal on, 18–19, 49
  industrial pollution in, 96–97
  isolation of, 7–8, 11, 38, 160
  management of, 78, 82–83, 99–100,
    103–4, 125, 295–96, 297–99, 305, 316,
    318
  and Mississippi River, 152, 271
  nutrients in, 217–18
  protections of, 274
  purposeful introduction of invasive spe-
    cies in, 88–92, 103–4
  recreation industry in, 84–85, 89, 94, 160
  regulating water withdrawals from,
    248–49
  scale of, xii–xiii, 138, 151
  self-healing, 316–17
  shoreline of, 7, 144, 208, 265, 266
  surface elevation of, 13, 14, 19
  veto power of state governors regarding,
    272–75
  water coursing through to the sea, 13–14,
    13, 38, 108, 151
  water evaporation on, 284–85, 291
  water held in public trust, 273
  water levels in, 277–78, 284–85, 288–99
  watershed borderline, 248–49, 253, 271,
    272
  water temperatures in, 73, 283–84, 291
Great Lakes Commission, 295, 297
Great Lakes Fishery Commission, 72,
    82–83, 318
"Great Lakes Restoration Initiative," 310–11
Great Lakes Water Wars (Annin), 266
Great Plains, irrigation plans for, 267–68
Green, Phyllis, 142–43, 147
Green Bay, Wisconsin, 190–91
Green River, Utah, 153
Griffiths, Ron, 119, 120
Gripentog, Bob, 193
Gulf of Mexico, 151, 152, 153, 162, 248, 254
Gulf of St. Lawrence, 8, 9
Gustafson, Tom, 249
Gustaveson, Wayne, 206–7, 211

Hall, Jay, 100–102
Hand, David, 143
Hanson, Dale, 320
Harrison, Benjamin, 14
Hawley, Jesse, 15–16
Hebert, Paul, 113–14
Helmick, Jon S., 31
Hemingway, Ernest, 78, 81, 90
Henderson, Henry, 310, 311
Henderson, Scott, 155

Henriksen, Charlie, 302-3
herring:
    lake, 39, 88
    management of, 77
    river, *see* alewives
Hickey, George, 34
Holey, Mark, 40-41
Holmes, Oliver Wendell Jr., 164
home septic systems, 234
Hoover Dam, 192, 194-95, 204-5
Hudson Bay, 248, 267, 280
Hudson River:
    and Erie Canal, 15-17
    sea lampreys in, 50
humpbacked peaclam, 109
hydroelectric power, 22, 280
hydrologic regime, *see* climate change

ice age:
    and fish, 36-37
    lakes shape-shifting in, 37-38, 265
    plants and animals in, 37
ichthyosaurs, 49
Idaho, invasive species in, 208
Illinois Department of Natural Resources,
    180, 181
Illinois River, 163, 164, 167, 171, 176
Indiana Dunes National Lakeshore, 217
Institute for Fisheries Research, 53
International Conference on Aquatic
    Invasive Species, 118-20
International Joint Commission, 293-96,
    297
invasive species:
    access via canals, 45, 48, 49, 50, 111,
        164-65, 189-90
    accidental discoveries of, 136, 137, 184
    in ballast water, xv-xvi, xvii, 35, 109, 134-
        36, 144-45, 147, 169, 188, 308-9, 311
    economic costs of, 147, 196-97, 309, 310,
        315
    effects of, xvi, xvii, 45, 123, 130, 135, 156-
        57, 308, 315
    equation factors of, 142
    as ever-growing national problem, 164,
        308
    in food chain, 49, 156-57
    genetic solutions to problems of, 315-16
    international conferences on, 118-20
    native predators vs., 317-21
    other pathways for, 144-45, 181
    pesticide use against, 138-39, 153-54
    purposeful introduction of, 88-92,
        103-4, 144
    on recreational boats, 137, 193-95, 198-
        208

research on, 168-69
and St. Lawrence Seaway, xviii, 109-14,
    135, 143, 144, 147, 165, 169, 187-88,
    191, 308-10
sleeper colonies of, 136
successful spread of, 179, 187-88
tallies of, 318
use of term, 135
zero tolerance of, 142, 311
Invasive Species Advisory Council, 168
Isle Royale National Park, 142-43, 147

James Bay, 267
Janesville (Wisconsin) *Daily Gazette*, 115-16,
    117
Janssen, John, 320
John Martin, 287, 288, 298
Johnson, Charles, 113
Joliet, Louis, 152, 162, 188
Jones, Alfred "Long," 262, 263
Jones and Laughlin, 116

Kenney, Pat, 35
Kernan, Lee, 95
Kew, Harry Wallace, 112
Keystone XL pipeline, 247
King, Louis, 62, 63
Kinsey, Alfred, 57
Klump, Norris, 232-34
Kohl, Robert, 24
Kompoltowicz, Keith, 296
Kortman, Cliff, 62-63
Koyen, Ken, 71-72, 300-303
Krumenaker, Bob, 281
Kuptz, Patric, 277-78

Lachine Canal, 11, 12-13
Lachine Rapids, 9-10, 11, 12, 22, 27
Lake Baikal, Siberia, 326
Lake Champlain, Vermont, 189-90
Lake Erie:
    alewives in, 67, 68
    algae slick on, xvii, 216, 225-31, 234-35,
        236, 238, 239, 296
    algae studies of, 239-40
    Asian carp in, 310
    birds killed on, 130-31
    and Detroit River, 234-35
    drinking water from, 232, 237-38, 239-
        42
    Dr. Seuss's mention of, 217, 222, 223-24
    elevation of, 14
    and Erie Canal, 15-18
    eutrophication of, 218, 221-22
    fish population in, 218
    freighters on, 109, 288

Lake Erie (*continued*)
and Great Black Swamp, 212–13, 214, 215–16, 225, 231, 233
and Gulf of Mexico, 152
HABs (harmful algae blooms) of, 239
isolation of, 7
Lake Huron connection to, 108–9, 222
lake trout in, 46, 61
microcystis in, 224–25, 229, 232, 235, 237, 244
Monroe Power Plant on, 120
Niagara River as outlet for, 14, 108, 222, 279
Nickel Plate Park beach, 166
nutrients in, 217–18, 229
phosphorus in, 220–22, 224–30, 234, 237, 243
quagga mussels in, 122
recovery of, 222–23, 232, 242–43
sea lampreys in, 51–52
and spring rains, 228, 229
as Walleye Capital of the World, 223
as waste receptacle, 216, 217, 221, 222, 234, 296
water levels of, 293
water retention time, 222, 232
and Welland Canal, 19–20, 51
western basin of, 228, 231, 234, 243
zebra mussels in, 119
Lake Huron:
alewives in, 67, 68, 74, 103, 127, 306
charter boats on, 304–8
and Chicago canal, 287
economic boom in, 102
fish population crashes on, 127, 304–5, 307
food chain in, 306
Georgian Bay study, 292–94
ice on, 296
inflow via St. Marys River, 14
isolation of, 7
Lake Erie connection to, 108–9, 222
Lake Michigan connection to, 14, 104, 107, 222
lake trout in, 46, 64, 84, 306, 320
mussel impacts on, 123
native species renaissance on, 308
and St. Clair River, 287–90, 295
salmon in, 100–103, 104, 105, 106, 304–7
scale model of, 290–91
sea lampreys in, 45, 46, 306
water diverted from, 287
water evaporation on, 284–85
water levels of, 279–80, 284, 285–86, 287–88, 289, 291–99

Lake Lanier, Georgia, 254
Lake Mead, Nevada, 192–96, 198, 203–4, 207
Lake Mendota, Wisconsin, 137–38
Lake Metonga, Wisconsin, 190, 191
Lake Michigan:
alewives in, 67, 68, 69–74, 75, 92, 105, 106, 126–27, 157, 319
and artificial fish "farming," 98
birds killed on, 130–31
and Chicago canal, 162–63, 162, 287
and Chicago River, 152, 162–63, 175
chinook salmon in, 95, 105–7
coho salmon in, 76, 91–94
commercial fishing on, 71, 100, 300–304, 303
drinking water from, 161–63
fish harvests in, 46
fish surveys of, 124–27, 184, 308
ice on, 296
isolation of, 7
Lake Huron connection to, 14, 104, 107, 222
lake trout in, 46, 64, 319–20
management of, 125, 303, 319
mussel impacts on, 123
native species in, 88, 100, 318
nutrients in, 105
and Pacific salmon, 85–87
and Platte River, 90
rainbow smelt introduced into, 88
and St. Clair River, 287–89, 295
salmon crash in, 102–3, 308
sea lampreys in, 45, 46, 52
Secchi depth of, 124
and sewage treatment systems, 161, 163–64, 183
sports fishing on, 125, 319
surface area of, 125–26
water diverted from, 163–64, 275–76, 287
water evaporation on, 284–85
water levels in, 278–79, 280, 284, 285–86, 287–88, 289, 291–99
water temperatures in, 284
whitefish in, 301–4, 306
Lake Ontario:
alewives in, 66–67, 68, 74
birds killed on, 130–31
cargo carried to, 12–13
mussel impacts on, 123
Niagara Falls as barrier, 10
outflow of other lakes into, 7, 14, 38, 108, 279
sea lampreys in, 50
water levels of, 279
and Welland Canal, 51

Lake Powell, Utah, 198–203, 200, 205–7, 211
"lakers," 188
Lake St. Clair, 108–9, 110, 111, 119, 187, 188,
    189
Lake Superior:
    alewives in, 67, 68
    balance of inflow and outflow, 13–14
    brook trout in, 142
    dynamic system of, 13
    headwaters flowing from, 13–14, 108
    ice on, 281–82, 283
    Isle Royale National Park in, 142–43, 147
    isolation of, 7, 123
    lake trout in, 61, 74, 84
    lampreys poisoned in, 64
    plans to tap for irrigation, 267–68
    plan to divert water to Asia, 272–74, 294
    sea lampreys in, 45, 64
    water diversion into, 280
    water levels in, 279
    water temperatures on, 283
lake trout, 38, 39–46, 40
    Applegate's view of, 74, 84–85
    classification of, 43–46
    collapse of stocks, 45, 301, 305
    different names for, 42
    effect of alewives on, 305–6
    evolution of, 41, 76
    harvests of, 44
    management of, 77, 83, 319
    as native species, 88, 317
    overfishing of, 42, 67
    restoration of, 54, 74, 84, 89, 95, 96, 98,
        305, 319–20
    and sea lampreys, 45–46, 52, 56, 61, 64
    size variations of, 41–42
    and sports fishing, 85, 319, 321
    thiamine deficiency in, 305
Las Vegas, 198, 270
Laurentian Great Lakes, xii
LeClair, Pete, 100
Legler, Nick, 106
Leopold, Aldo, 64, 316
Libben, Mike, 235–36
limnology, 137
Lin, S. Y., 155
Lincoln, Mary Todd, 261
Lodge, David, 168–69, 171–74, 176–78, 179,
    180, 181–83, 191
Loeffler, Steve, 233–34, 235

Mackie, Gerry, 119
Maclean, Norman, 81
MacNider, Hanford, 21
Madenjian, Chuck, 308
Madison, James, 16

Mahon, Andy, 171
Malaysia, fish introduced from, 154–55
Mao Zedong, 268–69
Marquette, Father Jacques, 152, 162, 188
Matych, Tom, 317–18
Maumee Bay State Park, 230
Maumee River, Ohio, 215, 227, 231, 235, 236,
    240, 243, 257
McClure, Andy, 238–40
McCormick, Gerald, 255
McElroy, James, 261–63
McLean, Malcolm Purcell, 25–27
McNally, Tom, 75, 76
Mediterranean Sea, 4–7, 17, 23, 24
Metropolitan Water District of Southern
    California, 196–97
Michigan Department of Conservation, 84
microbiology, 164
microcystis, 224–27, 229–31, 232, 235,
    237–38, 239–40, 244
Middlebrook, Craig, 144, 145, 210
Midwest Steamship Agency, 24
Mile Marker Zero, 251
Miller, Dick, 29
Miller, Robert Rush, 67
Milwaukee:
    and commerce, 19
    severe storms in, 296–97
minnows, 318
Mississippi River:
    Asian carp spread in, 176, 309
    drinking water from, 163
    water from Lake Michigan flowing into,
        163
    water levels of, 272
Mississippi River basin:
    area and scope of, 151–52
    Army Corps of Engineers study, 309–10
    carp released into, xvii, 156–61, 205, 313
    Chicago canal connection to Great Lakes,
        xvii, 122, 151–53, 162–64, 184, 188–89,
        271, 309
    and water diversions, 275
Missouri, lawsuit against Chicago canal,
    163–64
Missouri River, and Ogallala Aquifer, 268
Mitchell, Andrew, 154
mollusks, 38
    eaten by carp, 154, 310
    see also mussels
Monroe, Michigan, water supply, 120–21
Moore, Bryan, 203–4
Mormon Church, 42–44
moss piglets (tardigrades), 139–40
Moulinette, Canada, flooding of, 33–34
Murray-Darling basin, Australia, 313

Musgrave, Mark, 230–31
mussels:
    and botulism outbreaks, 130–31, 135
    economic costs of, 196–97, 202, 208, 211
    as "ecosystem engineers," 229, 231
    as filter feeders, 229
    in food chain, 39, 103, 128, 130, 146, 229,
        301, 302, 318
    shells of, *121*
    survey of, 123–27
    survival out of water, 191–92
    transport of, as felony, 198–203, 205,
        207
    uncontrollable, 196, *196*, 211
    veligers, 112, 188–89, 205
    see also quagga mussels; zebra mussels
Muter, Mary, 291–92

Nalepa, Tom, 118, 119
National Aeronautic and Atmospheric
    Administration, 270
National Aquatic Nuisance Species
    Clearinghouse, 147
National Oceanic and Atmospheric
    Administration (NOAA), 118, 229, 239,
    240, 284
National Park Service, 193, 198–99, 203–4,
    205–7
National Wildlife Federation, 97
Native Americans, attacks from, 11
native species, adaptation of, xviii, 302–4,
    305–8, 316–17, 320
*Natural History of the Sea Lamprey,
    Petromyzon marinus, in Michigan*
    (Applegate), 53–60
Natural Resources Defense Council, 66
Nature Conservancy, 178
navigational dredging, 288–89
navigation locks, 11–12, 16, 19, 24–25, 173,
    174–75, 178, 279
Netherlands, dikes in, 267
neurotoxins, 136
Nevada, invasive species in, 195, 199, 207
Newport State Park, Wisconsin, 128
New York City:
    and Erie Canal, 18–19
    water shortages in, 250, 271–72
Nguyen, Kathy, 254
Niagara Falls:
    bypassed by canals, 19–20
    cliffs of, 7, 10, 19
    erosion of, 7
    ice at, 28
    as impassable barrier, 10, 13, 14
    Niagara River flowing over, 108, 279
Niagara River, 14, 38, 108, 222, 279

Nicolet, Jean, xi, xii, xiii
ninespine stickleback, 38
nitrogen, 236
Nixon, Richard M., 117
Noah's Flood (Ryan and Pitman), 6–7
nonnative species, *see* invasive species
northern pike, 117
Northwest Airlines flight 2501, disappear-
    ance of (1950), xii–xiii
Northwest Indian War, 213
"not-my-fault" phenomenon, 236
Notre Dame, Environmental Change
    Institute, 168–69, 171–75, 177, 180
Nova Scotia, offloading cargo in, 145
nucleotides, 170–71
nutrients:
    and eutrophication, 218
    farm management of, 236
    in food chain, 217–18, 229

Obama, Barack, administration of, 176, 178,
    243, 247
Oberlin, David W., 33
Oberstar, James, 30, 135
Odom, Gary, 255
Ogallala Aquifer, 267–68, 274
Ohio:
    and fishery management, 83
    and phosphorus discharges, 243
    water pollution regulations in, 116–17
Ohio Environmental Protection Agency,
    240
oil tankers, double-hulled, xvii
Olin, Chauncey, 256–59, 275
O'Neill, Chuck, 147
Oregon Fish Commission, 76
Oregon Moist Pellet, 87
Orme, Tennessee, running out of water,
    253

Paczocha, Frank C., 52
Panama Canal, navigation locks of, 24–25
PCBs, 97
Peabody, John, 166–68, 171–75, 176, 177–78,
    179, 181–82, 185
perch, 39, 77, 88, 100, 301, 306, 317–21
Perdue, Sonny, 253
Perrier, Louis Eugene, 259
pesticides, to kill invasive species, 138–39,
    153–54
philosopher's stone, 220
phosphates, 220
phosphorus, 218–30
    and algae outbreaks, 231, 234–35
    and Cuyahoga fire, 232
    and farming practices, 228, 233, 236, 237

particulate vs. dissolved, 228, 236
reductions in, 232, 235, 242–43
testing soils for, 235
photosynthesis, 224
phytoplankton, 38, 124, 304
Pinchot, Gifford, 84
placoderms, 49
plankton:
eaten by alewives, 65, 103, 105
eaten by carp, xvii, 154, 156–57, 310
eaten by mussels, 120, 123, 128, 318
in food chain, 38–39, 41, 103, 113, 156–57,
207, 224, 318, 319
phytoplankton, 38, 124, 304
plummeting populations of, 103
surveys of, 123–27, 137
and Tanner's experiment, 80
zooplankton, 38, 73, 113, 139, 224, 305,
318, 319
Plant, Ernie, 307
Platte River, 90
plesiosaurs, 49
poison:
on Chicago canal, 175–78, 179–80, 182,
185–86
and daughterless gene, 312–15
and extinction, 314
invisible in water, 238
microcystin, 232, 239
phosphorus in, 220, 236
testing, 61–65, 83
toxic algae blooms, 216, 224–25
weed killer, 153–54
poison gas, 220
polar vortex, 296
Pollack, Lana, 294–95, 297–98
potassium, 236
potassium permanganate, 239
Potomac River, and canal, 14
predators, apex, 64
Prince William Sound, oil spill in, xvii

quagga mussels:
and algae, 227
discovery of, 122–23
ecological damage caused by, 122–24,
128, 130, 194, 211
in food chain, xviii, 130, 207, 229, 304
in freighters' ballast water, 131, 188, 308
gobies feeding on, 127, 128, 130
invasive colonies of, xvi, xvii–xviii, 122,
130, 135, 207, 301, 318
and pleasure boats, 193–95, 200–203, 206
spread of, 187–89, 191, 193–201, 203–6,
208, 211, 229
Quinn, Frank, 285, 289–91

railroads:
competition of, 30, 32
off-loaded cargo shipped via, 145
rain:
and farming, 228, 229, 237
lack of, 285
"100 year storms," 296
phosphorus in, 220
Ranger III, 142–43
razor-toothed snakehead, 165
Reames, Tony, 253
recreational boats:
decontamination of, 199–204, 207
invasive species on, 137, 193–95, 198–208
mandatory inspections of, 199–202, 205
quarantining of, 204
recreational fishing, *see* sports fishing
"Redneck Fishing Tournament," 158
Red Sea, and Suez Canal, 24
Regelman, Dwight, 200–202, 210
Republic Steel, 116–17
Revolutionary War, 11
Richardson, Bill, 269–70
River Runs Through It, A (Maclean), 81
Rocky Mountains, as barrier, 192, 268
Roebber, Paul, 284, 284–285, 297–99
Rogner, John, 176, 177, 180
rotifers, 224
Russian River, California, 153

St. Clair River, 14, 279–80, 286–96, 297
St. Lambert Lock, Montreal, 144–45
St. Lawrence River:
canal running parallel to, 11–12
dam system on, 279
Great Lakes outflow into, 7, 13, 13, 38, 151,
279
invasive species in, 121
Lachine Canal finished on, 12–13
Lachine Rapids on, 9–10, 11, 22, 27
navigation channel of, 3–4, 288
navigation locks of, 19–20
villages flooded along, 33–34
winter shutdowns of, 20, 32, 33
St. Lawrence Seaway:
civil fines in, 210–11
commerce in, 30–33
completion of, 27
construction of, 22, 23–25, 33–35, 109
debt of, 30
digging started on, 11
freighters on, xv, 21, 30, 109, 113, 145–47,
187–88, 191, 208–11
as Great Lakes "front door," xvii, xviii, 153,
309
and hydropower dam, 22

St. Lawrence Seaway (*continued*)
  idea of, 3–4, 8, 20–22
  joint U.S.–Canada ownership, 310
  navigation locks of, 24, 27, 29, 144–45,
    208–11
  open to invasive species, xviii, 109–14,
    135, 143, 144, 147, 165, 169, 187–88,
    191, 308–10
  optimistic predictions for, 23–24, 30
  overseas trade in, 31, 198, 309
  as regional navigation corridor, 31
  salinity tests in, 209–10
  shortcomings of, 28–29
  water flowing to Atlantic Ocean via, 108
  and winter conditions, 28, 31–32, 33
St. Lawrence Valley, 38
St. Louis, typhoid in, 163–64
St. Marys River, 13–14, 279
salmon, 39–40, 66
  artificial food for hatcheries, 87
  artificially stocked, 77–78, 86–100,
    102–4, 135, 305, 306, 316, 318
  Atlantic, 86
  chinook, 95, 98–99, 101–3, 104–7, 304–7,
    320
  coho, *see* coho (silver) salmon
  and DDT, 97
  diseases of, 99
  experimental breeding of, 98–99
  and food chain, 67, 304–5, 307, 319
  hatchery-raised, 104–7, 125, 197
  overstocking of, 99, 100
  Pacific, 85–87, 95, 100, 197, 305, 308
  population crash, 100–103, 105, 304–5,
    307–8
  restoration of, 197
  "salmon craze," 93–94, 97, 101, 307
  as sport fish, 126, 304, 320, 321
  unsustainable numbers of, 102–3
  wild vs. stocked, 102–3
salmon boats, 94
"salties," 145–46
Sandburg, Carl, 161
Sand County Almanac, A (Leopold), 64
San Francisco Bay, invasive species in, 144
Santavy, Sonya, 108–12, 114, 118, 119, 136
Santiago, 286
Scavia, Don, 243
schooners, 20, 288
sculpins, 38, 125
sea lampreys, 38, 45–65, 50, 123
  as anadromous, 49–50
  Applegate's research on, 53–61, 223
  in Atlantic Ocean, 49
  control strategies, 60–64, 82, 83, 223, 311,
    319

escape instinct of, 55
first discovery in Great Lakes, 50
and fishery collapse, 45, 65
and fish-killing virus, 165
fossil of, 46–49
humans attacked by, 46
as invasive species, 135, 306, 316
killer stage of, 54–55
in Lake Erie, 51–52
in Lake Michigan, 45, 46, 52
and lake trout, 45–46, 52, 56, 61, 64
migration of, 51, 53, 54, 56, 59–60, 76
mouth of, 46, 47, 48, 55, 56, 59, 59
as prey, 49
spawning, 55, 56–58
spread of, 45–46, 50–52
survival of, 48–49
transformation of, 58–59
Sears, Richard W., 261
Secchi disk, 123–24
Seneca Chief, 17
Serageldin, Ismail, xiii
Settele, Brian, 320–21
Seuss, Dr. (Geisel), 243
  The Lorax, 217, 222–24
sewage systems:
  Chicago/Lake Michigan, 161–66, 183,
    189, 261, 271, 296
  Great Lakes as receptacles for, 222, 234,
    248
  and polar vortex, 296
  treatment experiments, 155–56
shad, 85, 207
shifting baseline phenomenon, 130
shipping industry:
  and Clean Water Act, 141, 143
  containerships, 26–27, 32–33
  costs of, 146–47
  domestic commerce, 146–47
  dumping ballast water, *see* ballast water
  and erosion, 294
  freighters in, *see* freighters
  huge business of, 30–33, 145–47
  and invasive species, 308–9
  overseas trade, 31, 145–46, 191, 198, 309
  pilot regulations, 29
  and riverbed mining, 288–89, 294
  seasonal closure in, 31–32
  served by Army Corps of Engineers, 290,
    309–10
Silent Spring (Carson), 153
Sinclair, Ralph, 113
Sleeping Bear Dunes, 128
smelt, rainbow, 88
Smith, Henry, 45
Smith, Joseph, 42, 44

Snake River, 197
snowflakes, 36
soap, phosphorus in, 221, 222, 236
Southern Nevada Water System, 195
spiny water fleas, xvii, 113, 135, 137
sports fishing:
    bigger boats for, 94
    charter boats, 304–8, 317, 321
    and DDT, 97
    economic costs of invasive species to, 147
    economic effects of, 94, 96, 102, 319, 404
    and fishery management, 81, 84, 104,
        107, 125, 305, 316, 319–20
    just for fun, 78–79, 83, 89, 317
    methods used in, 93
    re-emergence of, 304
    salmon crash, 100–103, 304
    "salmon craze" in, 93–94, 97, 101, 307
    salmon stocked for, 77–78, 89, 90–94,
        96–100, 102–4, 107
    and small craft advisory, 94
steelhead, 117
Strait of Gibraltar, 4, 5–7
Straits of Mackinac, 14, 280
Strang, James J., 42–44
Strawberry Creek, 105–6
Strickler, Rudi, 139, 140
Stupak, Bart, 273
sturgeon, 39
Sturgeon Bay Canal, 105
Suez Canal, global commerce in, 24
Supreme Court, U.S.:
    and Chicago canal, 163–64, 271, 280
    and state boundary disputes, 255, 256
Swigert, Fred, 132–33
Szylvian, Kristin M., 77, 89, 98

Tanner, Howard, 54, 74, 78–87
    and alewife die-off, 83–84, 98
    in Army Signal Corps, 88–89
    coho salmon stocked by, 86–94, 96–100,
        103–4
    experiments of, 80, 89–90, 98–99, 306,
        308, 317, 320
    move to Michigan, 82–83
    and sports fishing, 83, 84–85, 89, 94, 97,
        107
tardigrades, 139–40
Tennessee:
    statehood of, 250
    state line (35th parallel) of, 250–56
Tennessee River, 254, 255
Thorbahn, Richard, 236–37
Three Mile Island, 268
threespine stickleback, xvii
Thresher, Ron, 313–15

Thunder Bay, 143
tilapia, farm-raised, 197
Titanic, 133
Tocqueville, Alexis de, 216–17
Tody, Wayne, 85–86, 89, 92, 93, 96, 103–4,
    306
Toledo, Ohio:
    drinking water, 230–31, 238–40, 244
    "Fortress of the Lake" water intake, 239
    and microcystin levels, 244
    as warning, 243–44
Toyota Motor Corporation, 31
transportation, evolution of, 20
Treaty of Greenville (1795), 213
trout:
    brown, 306, 320, 321
    coaster brook, 142
    cutthroat, 81
    hatchery-raised, 125, 135
    lake, *see* lake trout
    as sport fish, 81
typhoid, 163–64, 262, 263

United States:
    and St. Lawrence Seaway construction, 25
    and shipping industry, *see* shipping
        industry
    western expansion of, 14–15
University of Notre Dame, 147, 168–69,
    171–75, 177, 180, 210
U.S. Bureau of Commercial Fisheries, 96
U.S. Bureau of Fisheries, 88
U.S. Bureau of Reclamation, 194, 195, 204
U.S. Department of Commerce, 88
U.S. Department of Interior, 116, 154–55
U.S. Department of Transportation, 131
U.S. Fish and Wildlife Service, 60, 62, 88,
    176, 180, 182, 311
U.S. Fish Commission, 88
U.S. Geological Survey, 125–27
U.S. Seaway, 30, 32, 135, 144
U.S. Steel, 116–17
Utah:
    boats quarantined in, 202
    invasive species in, 195, 207
    mussel possession illegal in, 198, 201–2
Utah Division of Wildlife, 206
Utica, New York, 18

Vander Zanden, Jake, 137, 187
Van Herick, Russ, 315–16
veligers, 112, 188–89, 205
Virginian, The, 199

walleye, 39, 223, 306, 307, 317
Washington, George, 12, 14–15

Washington Island, 300–301
Washington state, felony cases in, 207–8
water:
    access to, 249–50
    artificial purification systems, 232
    biologically contaminated, xiv–xvi, 35,
        109, 111, 131, 141, 198, 225
    chemically contaminated, 115–17, 217
    chlorine in, 239
    and climate change, see climate change
    diversion to Asia, 272–74
    DNA sampling of, see DNA
    evaporation-precipitation cycles, 279,
        284–86, 291–98
    in freighter ballast, see ballast water
    freshwater, xi, xii, 82, 248, 249, 265
    groundwater reserves, 270
    invisible toxins in, 238
    manmade channels, 215
    natural purification of, 231
    and nutrient management, 236
    nutrients filtered out by invasive species,
        122–23
    public supply systems, 225, 232, 252–56,
        261–62, 270–72
    radium in, 264
    retention time, 222
    saltwater, 265
    and sewage treatment systems, 155–56
    shortages of, 250, 253–55, 270–71
    state vs. federal regulations on, 116–17
    storm runoff, 216
    WHO standards for, 237
    withdrawals from Great Lakes, 248–76
    see also Clean Water Act
water bears (tardigrades), 139–40
water fleas, xvii, 110, 113, 135, 137, 224
water raids, 256, 263
watersheds, 252
Waukesha, Wisconsin, 256–64
    artesian springs in, 259, 260, 261, 264,
        274
    mineral water from, 259–62
    radium in water of, 264
    water diverted from Lake Michigan to,
        275–76
    water supply, 262, 263–64
    water wars in, 263, 275
Waukesha County, Wisconsin, 249, 250,
    259
Wayne, "Mad" Anthony, 213
weed killer, 153–54

weirs, 60–61
Welland Canal:
    expansions of, 20, 49, 51
    flow of water in, 51
    "lakers" on, 188
    navigation locks in, 19–20, 25
    Niagara Falls bypassed by, 19, 38
    opening of, 51
    traffic jams in, 28
    winter shutdowns of, 20
Western Electric Company, 131
West Lost Lake, 80, 90
whelk, veined rapa, 144
whitefish, 39, 41, 45, 77, 83, 88, 100, 301–4,
    306
Who Killed Lake Erie? (documentary), 222
Willett, Leonard, 194–96
Wisconsin, inland lakes of, 190–91
Wisconsin Department of Natural
        Resources, 95, 106
Woldt, Aaron, 185
Wooley, Charlie, 182
World Health Organization (WHO), 237
World War II, incendiary bombs in, 220
worms, tubificid, 110, 113
Wyoming, mussel possession illegal in, 198

Yangtze River, 269
Yellowstone River, 81
Young, Brigham, 42, 43

zebra mussels:
    cost of, 120–21, 122
    discovery of, 109, 110, 136
    Dreissena polymorpha, 110
    and food chain, 120, 121, 130, 229, 310
    in freighters' ballast water, 111–12, 118,
        131, 188, 308
    gobies feeding on, 127, 128
    invasive colonies of, xvi, 110–13, 130, 135,
        301
    mandatory boat decontamination for,
        200–202
    in native fish diet, xviii
    plankton eaten by, 120, 128
    reproduction of, 112
    selective filtering by, 229
    spread of, 119–23, 187–91, 198–200, 208
    tethers of, 120
zooplankton, 38, 73, 113, 139, 224, 305, 318,
    319
Zuiderzee, Netherlands, 267